Climatology
for Airline Pilots

Books on the JAR syllabus

Ground Studies for Pilots
Radio Aids
Sixth Edition
Roy Underdown & David Cockburn
0-632-05573-1

Ground Studies for Pilots
Navigation
Sixth Edition
Roy Underdown & Tony Palmer
0-632-05333-X

Ground Studies for Pilots
Plotting and Flight Planning
Sixth Edition
Roy Underdown & Anthony Stevens
0-632-05939-7

Ground Studies for Pilots
Meteorology
Third Edition
Roy Underdown & Anthony Stevens
0-632-03751-2

Climatology for Airline Pilots
Roy Quantick
0-632-05295-3

Human Performance and Limitations in Aviation
Second Edition
R.D. Campbell & M. Bagshaw
0-632-04986-3

Aircraft Performance Theory for Pilots
P.J. Swatton
0-632-05569-3

Aviation Law for Pilots
Tenth Edition
Roy Underdown & Tony Palmer
0-632-05335-6

Climatology
for Airline Pilots

H.R. Quantick FRAeS, FRMetS

b

**Blackwell
Science**

© 2001 by
Blackwell Science Ltd
Editorial Offices:
Osney Mead, Oxford OX2 0EL
25 John Street, London WC1N 2BS
23 Ainslie Place, Edinburgh EH3 6AJ
350 Main Street, Malden
  MA 02148 5018, USA
54 University Street, Carlton
  Victoria 3053, Australia
10, rue Casimir Delavigne
  75006 Paris, France

Other Editorial Offices:

Blackwell Wissenschafts-Verlag GmbH
Kurfürstendamm 57
10707 Berlin, Germany

Blackwell Science KK
MG Kodenmacho Building
7–10 Kodenmacho Nihombashi
Chuo-ku, Tokyo 104, Japan

Iowa State University Press
A Blackwell Science Company
2121 S. State Avenue
Ames, Iowa 50014-8300, USA

First Edition published 2001

Set in 10/13pt Times
byDP Photosetting, Aylesbury, Bucks
Printed and bound in Great Britain
at the Alden Press Ltd, Oxford and Northampton

The Blackwell Science logo is a trade mark of Blackwell
Science Ltd, registered at the United Kingdom Trade
Marks Registry

DISTRIBUTORS
  Marston Book Services Ltd
  PO Box 269
  Abingdon
  Oxon OX14 4YN
  (*Orders:* Tel: 01235 465500
          Fax: 01235 465555)

USA
  Blackwell Science, Inc.
  Commerce Place
  350 Main Street
  Malden, MA 02148 5018
  (*Orders:* Tel: 800 759 6102
          781 388 8250
        Fax: 781 388 8255)

Canada
  Login Brothers Book Company
  324 Saulteaux Crescent
  Winnipeg, Manitoba R3J 3T2
  (*Orders:* Tel: 204 837-2987
          Fax: 204 837-3116)

Australia
  Blackwell Science Pty Ltd
  54 University Street
  Carlton, Victoria 3053
  (*Orders:* Tel: 3 9347 0300
          Fax: 3 9347 5001)

A catalogue record for this title is available from the
British Library

ISBN 0-632-05295-3

Library of Congress
Cataloging-in-Publication Data
is available

For further information on
Blackwell Science, visit our website:
www.blackwell-science.com

Contents

Preface

What is climatology? The word 'climatology' is a derivation from the Greek word *Klimo* which means an incline, or slope. This refers to the angle of incidence of the sun's rays striking a particular place. This in turn affects the climate of that particular place. Climate considers the parameters of temperature, pressure, humidity, wind and precipitation.

There are many books written about climatology, but to date almost none on aviation climatology. This book will serve as a base-line reference for the airline pilot. It will also serve students of aviation climatology, and more than covers the JAR (Joint Aviation Rules) learning objectives for the new European professional pilots licences.

Students of aviation climatology in former times could probably only identify the weather one could expect along the 'Empire routes'. Nowadays, aeroplanes fly considerably higher, and are equipped with very much more sophisticated navigation, fixing and communications systems. Aeroplanes can now fly long distances non-stop and this, combined with the development of automated reporting systems on aircraft transmitted via satellite communications, enables new routes to be flown, for example across the Pacific and the Polar regions.

It is important for the airline pilot to understand the structure of the atmosphere as this is essential for the safety of passengers and crew. A knowledge of how the various seasons, which differ around the world, alter the atmospheric processes is also an important part of this understanding.

Traditionally, climatology has been concerned with the collection of statistics which express the average state of the atmosphere. Unfortunately, some of these statistics have little relevance to the natural environment as experienced by man. Available references are more often orientated to physical, biological and cultural environments. In this book attention has therefore been given to those aspects of climatology that are important in modern aviation, including temperature, precipitation, solar radiation, winds, upper winds and regional climatic environments in different parts of the world. Also included are particular local meteorological phenomena that affect flying operations.

To understand the climatic environment, it is necessary to study climate on a variety of scales, varying from the local to the global. This book examines the factors which determine mesoscale and macroscale climates, and the nature of the major climatic regions of the Earth, which basically means a study of the general circulation of the global atmosphere. Climatology is therefore better described as the mean physical state of the atmosphere, together with its statistical variations, both spatially and temporally, as reflected in the weather behaviour observed over a period of many years.

In more recent years, climatological variations have been observed in greater detail and in near real time by sophisticated global monitors, especially satellite sensors. Valuable information can now be obtained on upper winds and temperature for the high altitudes flown by aircraft. The advent of an automated reporting system giving such reports every few minutes can provide very detailed information for climatological studies.

The important aspects of the atmosphere affecting the flight of an aircraft are the location and nature of jet streams, areas of turbulence, location of storm clouds, and the low-level weather for safe landing and take-off. These features of the weather are the result of dynamic and thermal dynamic energy processes within the atmosphere, an understanding of which is essential for the pilot.

In order to appreciate the climatological implications, a reasonable knowledge of geography is required, with particular emphasis on mountain ranges and their alignment, so access to an atlas would be an advantage.

Some less frequently used terms are included in the text, and a glossary is included to complete the explanation. Two appendices detail local winds and briefly summarise weather at selected destinations.

Acknowledgements

I am grateful to many people and organisations for advice, facilities and assistance in gathering the information for this book. In particular.

Iain F. Kerr, Manager, Aviation Rules, CAA of New Zealand.
Kevin Chown, Editor, Remote Imaging Group Journal (RIG), UK.
Jacques Pougnet FRmetS, Secretary of the Mauritius Meteorological Society.
Peter Lindsley, CBM Publishing.
Les Hamilton PhD, RIG Committee member.
John Coppens, University of Cordoba, Argentina.
The Royal Meteorological Society of England.
European Space Agency (ESA).
The UK Meteorological Office.
Peter Wakelin, RIG Committee member.
US National Oceanographic and Atmospheric Administration (NOAA).
Satellite Receiving Station, Dept of A.P.E.M.E., University of Dundee, Scotland.
Dave Cawley, Timestep England (weather satellite equipment).
David Binks, Cranfield University.
HRQ Jan 2001

Introduction

In recent years, considerable changes have taken place in communications systems, and the interchange of meteorological information is no exception. Considering the nature of long haul aviation, pilots need forecasts of the main meteorological phenomena that is required for planning the flight. They also need to understand upper winds, temperatures, tropopause heights, jet streams, mountain waves, thunderstorm activity, tropical cyclones, clear air turbulence (CAT), volcanic activity and such phenomena when conducting the flight. Also, there is the terminal weather (TAFs – Terminal Air Forecasts) and the airports nominated as alternates, both en route and the destination.

Global weather forecasting is becoming a reality. The UK Meteorological Office (MO) is developing its Numerical Weather Prediction (NWP) model, and the resolution of the areas (grid squares) around the world is improving. This means an enormous amount of computer processing is involved, but so to is the acquisition of information.

The World Area Forecast Centres (WAFCs) under the provision of ICAO, is centred at two locations, the UK Met Office (Bracknell) and also Washington USA (based in Kansas City). Three INTELSAT 604 satellites provide global coverage. The UK Met Office uses one at 60° E (SADIS Satellite), and covers Europe, the Middle East and South Asia. The USA covers the other half of the globe. The satellites are in geostationary orbit.

The MO produces charts of significant weather from Flight Level 100 to Flight Level 450 for Europe and FL 450 to FL 630 for the North Atlantic. Also spot wind charts for the same areas. Significant weather includes jet streams, heights, direction and core speeds. The significant weather charts and associated spot winds are produced from FL 250 to FL 450 for the Middle East and South Asia.

Upper wind and temperature charts are produced for ten global regions, twice a day at nine levels. Thus, the total output is 396 charts a day. Only the significant weather charts are combined manually, the rest, $\cong 360$, are produced by automation.

The distribution of such charts presently is by the T4 FAX standard of 64 kbit/sec, but a new format to be used is 'GRIB' binary. This is more suitable for transmission of Grid Point Format charts. The GRIB code is contained in

1

WMO Publication No: 306, *Manual on Codes*, Volume 1.2, Part B – Binary Codes. The GRIB format will allow world atmosphere models to be transmitted, allowing airlines to optimise their tracks.

The MO increasingly relies on meteorological satellites to provide weather observations and nephanalysis particularly over the oceans. Aircraft will provide additional data, but the system will be automated. British Airways will have over 60 aircraft supplying fully automated weather reports. On average, the MO will receive 160 wind and temperature reports daily from each operational aircraft and these are used directly in producing the NWP forecasts, which are becoming the primary method of weather forecasting. This is done by solving a set of equations. A computer model of the atmosphere shows how weather conditions will change over time.

A valuable source of meteorological and climate observations is becoming available from the new Quickscat satellite – on board is NASA's SeaWinds instrument. Access to daily wind data and animations from the ocean-wind tracker are managed by NASA's Jet Propulsion Laboratory (JPL), Pasadena, California.

The heart of SeaWinds is a specially designed spaceborne radar instrument called a scatterometer. The radar operates at a microwave frequency that penetrates clouds. This, coupled with the satellite's polar orbit, makes the wind systems over the entire world's oceans visible on a daily basis. The measurements provide detailed information about ocean winds, waves, currents, polar ice features and other phenomena, for the benefit of meteorologists and climatologists. This data will be used operationally by forecasters and for numerical weather prediction models. Upper air observations are also obtained from suitably equipped ships on the Atlantic shipping lanes. This system is presently becoming operational. The MO will receive weather data twice a day for approximately 20 days of each voyage.

The MO recently (1998) stopped using the existing regional model for forecasting, and started to use the NWP model. In 1999 a new method of analysing the weather observations more suited to the NWP model was implemented. It is also more suited to the satellite data received. This and the new instruments on board the latest weather satellites are improving the performance of the NWP model, which in time will cover the globe. The computer processing required by the new model demands the use of super computers, i.e. the Cray T3E which can handle the changes to the NWP just mentioned.

Aircraft fitted with the ACARS (Aircraft Communications and Reporting System) Teleprinter system already receive Aircraft Operational Control (AOC), Airline Administrative Control (AAC) and Air Traffic Control (ATC). The system is an air to ground data link system used on HF, INMARSAT, and particularly VHF; however, HF, VHF and UHF frequencies are used. The cockpit equipment consists of a small printer, although, if this fails, a read-out can be seen on the alphanumeric display on the control unit. Through this system, pilots can be alerted to anything unusual which affects the current flight segment. This may include changing weather conditions, updating of TAFs, SIGMETs or mechanical information.

Long-haul pilots can look forward to a continuing improvement in the provision of meteorological information, which is approaching 'real-time' with the growing sophistication of the communications systems.

Part 1
Global Weather

Chapter 1

Global Air Circulation

1.1 Idealised circulation

The term 'idealised circulation' is used when describing global wind patterns resulting from atmospheric temperature and pressure gradients set up by the differentiation of solar radiation that is received at the Earth's surface.

The atmospheric circulation that is observed is very complex, and exhibits wide variations on both temporal and spatial scales. However, before looking at actual circulation patterns, it would be easier to look first of all at a very simplified model based on a planet that is not rotating, has a uniform homogeneous surface, and is in orbit around the sun. Solar radiation passes through the Earth's atmosphere and heats up the surface more strongly at the equator compared with higher latitudes towards the poles. Air would rise above the equatorial warmer region, and since pressure decreases less rapidly with height in a warm air column, pressure would therefore be greater at the higher levels than at higher latitudes (level for level) thus establishing a pressure gradient which sees the air flowing poleward. Continuity would be satisfied by the air descending at the poles and returning to equatorial regions at the surface, i.e. flow at the surface is from a high pressure area (descending air) to a low pressure area (ascending air at the warmer equatorial region). See Fig. 1.1.

Of course the circulation just described is not observed, because for one thing the earth is rotating, and secondly, when rotation is introduced the effect of the Coriolis force must be taken into account. Poleward movement of air will be deflected towards the east. This does take place, but at upper levels the circulation is limited to a maximum of about 30° of latitude before the air cools and subsides. Therefore, in each hemisphere on the rotating model we find a simple thermal circulation. This circulation is called a Hadley cell. Its limits lie between the equator and about 30° North and South latitude.

At the surface, belts of high pressure show where the poleward limits of Hadley cells (descending air) exist (subtropical high pressure belts). The surface winds have a marked easterly component (the trade winds) and blow from these high pressure areas towards the Equatorial trough (low pressure area).

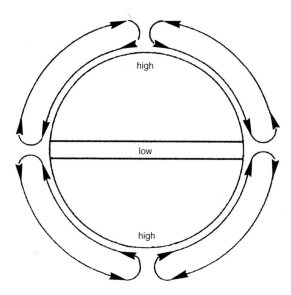

Fig 1.1 Air circulation (convective cell distribution) over a homogenous surface, non-rotating earth. Sometimes described as meridional circulation.

See Fig. 1.2 for a vertical profile of the air circulation and corresponding surface pressure area, and Fig. 1.3 for the surface wind directions.

The retention of air in the polar areas is observed. Polar air is very cold, and therefore more dense in contact with the cold surface, consequently high pressure must predominate, and an equatorial flow from high pressure to low pressure areas is set up.

Looking at the mid-latitudes, we find that between the subtropical high pressure areas and the polar high pressure, there is a band of low pressure in which surface outflows from the high pressure areas converge. The outflow from the polar highs moving towards equatorial regions is deflected towards the west, and the poleward flow from the subtropical highs is deflected towards the east. This does not contradict the deflection rule set by the Coriolis force, as deflection is to the right (in the northern hemisphere) from the *initial track*, and to the left in the southern hemisphere from the *initial track*.

In maintaining continuity, there must be ascent of air within this region and divergence aloft both polewards and equatorwards. These longitudinal (or meridional) components of flow are in fact quite small, but the dominant flow between the subtropical highs and about 60°N is towards the east. This region is referred to as the region of mid-latitude westerlies or variable westerlies or Ferrel westerlies. See Fig. 1.3 for the resultant surface wind directions.

To summarise, air flows in each hemisphere can be represented by three autonomous but collaborative cells, the Hadley cell in low latitudes, the Ferrel cell in mid-latitudes and a polar cell usually above about 60° pole-

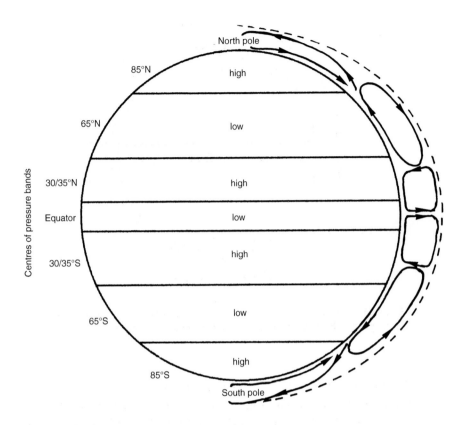

Fig. 1.2 For a rotating Earth, the simple air currents would be deflected by the Coriolis force. The circulation is disrupted, but the end result is still to *endeavour* to transfer air from equatorial regions to the poles aloft and an equatorial drift from the poles at the surface.

wards. Figure 1.4 shows in schematic form the main components, and Fig. 1.5 the three main cells.

Departures from the idealised circulation

The real earth is not homogeneous, and the pressure and temperature distributions are greatly affected by land/sea areas. There is more land surface in the northern hemisphere and mountain barriers inhibit air mass migration or block it. Seasonal variations are very marked, particularly in the northern hemisphere.

A further consideration must be taken into account and this is the axial tilt of the rotating Earth. If no tilt was observed, and the axis of tilt were perpendicular to the orbital plane, then there would be no change to the air circulation, and its pattern would remain unchanged throughout the year. The cells would remain in a fixed position throughout the year.

However, this is not the case and the Earth's tilt is about 23° 27′ (the

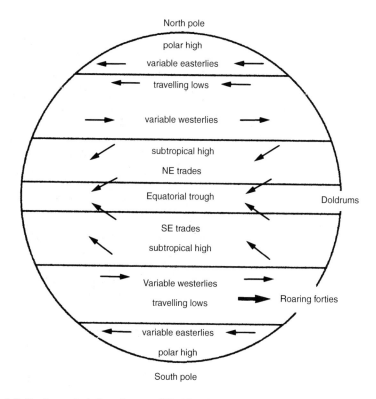

Fig. 1.3 Surface wind direction modified by the Earth's rotation.

obliquity of the ecliptic). As the Earth travels along its orbit, the meteor-ological equator (sometimes called the thermal equator) migrates north and south to the limits of 23° 27′ North and South latitude. The consequent effect of this movement is to displace the cells and associated pressure systems.

Over our homogeneous Earth's surface, we can expect the equatorial low pressure belt to move north and south within the above limits, that is to say in December it is at the maximum southerly limit (December solstice), and in June it is at the maximum northerly limit (June solstice). With the real Earth, these annual displacements are further modified by the land/sea configuration; however, the above description introduces the concept of *seasons*.

Temperature distribution

It is a fair statement to say that temperature decreases away from the equator to the poles, but there are quite large differences to this general rule. The main points to be noted are as follows.

(1)   Warmest temperatures occur over land in equatorial regions between positions 20°–30°N in July and 10°–20°S in January.

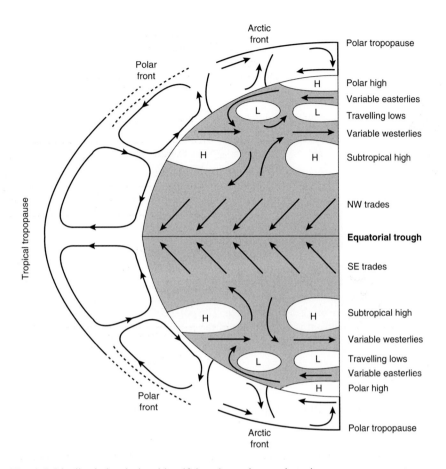

Fig. 1.4 Idealised circulation identifying the main quasi-stationary pressure systems. Note the position of the polar front, which is the poleward limit of a Ferrel cell, and the Arctic front which is the 'boundary' of a polar cell.

(2)   Coldest temperatures occur over land in the northern hemisphere in winter and it is generally colder in Siberia than at the North Pole.

(3)   In general sea temperatures will be warmer in winter (mid-latitudes) than in summer in the northern hemisphere.

(4)   Seasonal change of mean temperature may be as great as 60°C in Siberia, Canada or Greenland. At the equator, particularly over the sea, seasonal change is less than 5°C.

(5)   In North America, the Rockies protect the land mass from relatively cool moist prevailing westerlies. The absence of a barrier east–west allows full freedom of movement from the Caribbean northwards or the Arctic Circle southwards. In Europe there is no barrier to prevent east–west flow but the Alps do help protect the Mediterranean area from extreme cold weather from the north.

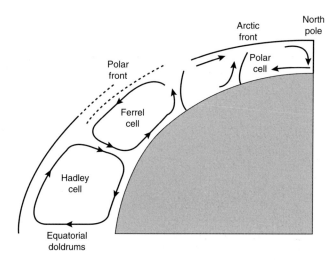

Fig. 1.5 Simplified cross-section through the atmospheric circulation in the northern hemisphere showing the location of the Hadley cell and the Ferrel cell. The subtropical anticyclone lies at 30° North. The polar cell lies to the north of about 60° latitude. The Arctic front is to the north of the polar front, but its position is less precise and is left out of this figure.

(6)   In Asia the Himalayas prevent southward spread of cold air from the Siberian high to India and Pakistan.

Pressure distribution

Look at the average pressure charts for January and July in Figs 1.6 and 1.7 and note how they compare with the ideal distribution in Fig. 1.4. The main features can be summarised as follows.

(1)   January – equatorial low pressure belt centred between 10° and 20°S with extensions further south over the land masses of Africa, South America and Australia.
(2)   July – equatorial low pressure belt centred near the geographic Equator with extensions northwards over Africa and particularly Asia.

Northern hemisphere

(1)   In January over the oceans the subtropical high appears between 20° and 40°N but extends over North America and Asia to 45°N.
(2)   The polar front is influenced by the high pressure systems and appears in sections, one over the Atlantic and another over the Pacific. The continuous formation and movement of depressions along these main frontal surfaces leads to the average pressures in the proximity of Iceland and the

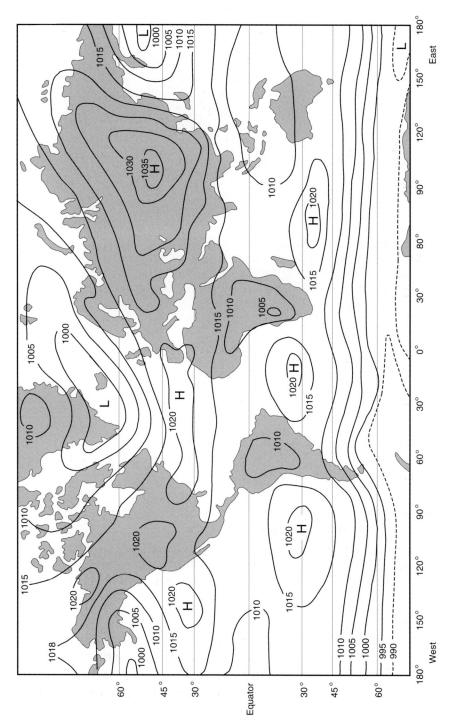

Fig. 1.6 Sea level pressure distribution (mb) in January.

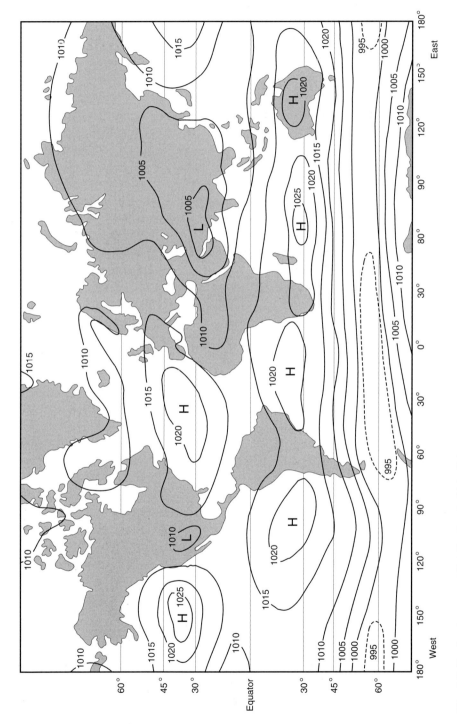

Fig. 1.7 Sea level pressure distribution (mb) in July.

Aleutians being low. On seasonal charts these are frequently termed the 'Icelandic and Aleutian' lows.

(3)   The pressure in the North Pole area is not quite as depicted in the ideal situation due to high pressure being found over the land masses of Asia and North America during January.

Southern hemisphere

(1)   Subtropical high – a well defined zone between 20° and 30°S over both land and sea areas.

(2)   Polar front – still approximates to 60°S.

Variations of pressure

In tropical areas pressures remain fairly constant and the major variations usually occur with tropical cyclones. Towards the poles, however, average pressure cannot be taken as actual pressures on any particular day due to system fluctuations. The best example is the 'Icelandic low', where it is possible to find an anticyclone over the region, but this does not negate the fact that the area has average low pressure.

Chapter 2
The Global Overview – Notes

2.1 Introduction

In this chapter, further details are introduced within the global atmospheric circulation. These occupy the macro scale, and affect large areas of the Earth's surface. Atmospheric heat exchange processes are discussed. The fundamental energy exchange meridionally from the warmer latitudes to the colder latitudes is effected by atmospheric pressure differences arising from heating of the atmosphere. Latent heat is another form of energy transference when condensation occurs in rising air currents. The oceans too transport heat energy from the warmer latitudes to colder latitudes. Heat is lost from the Earth by long wave radiation. The heat input by short wave radiation from the sun is balanced by the heat loss, but not uniformly over the Earth's surface. The term 'albedo' is introduced which is the relationship of the incident radiation to the amount re-radiated (reflected) by the surface.

Upper air waves (Rossby waves) are described and the correlation with surface weather phenomena. The ocean currents make their contribution to climate and the distinction between *warm water coasts* and *cold water coasts* are described. The term 'aridity' is used to describe certain areas in the world, and aridity is found over land and sea. Some general notes on desert areas are included, but specific weather phenomena in deserts are described in the appropriate geographical chapter.

Classic climatological phenomena are the 'trade winds'. Sailors have known about trade winds for hundreds of years, hence their name. They were also aware of the doldrums, which were not so beneficial. The mechanisms that produce the trade winds and the trade wind inversion are described.

Pilots do their best to avoid flying into active thunderstorms. Knowledge of their existence and extent are usually available at their briefing. However, active thunderstorms may be encountered unexpectedly and appropriate action taken in avoiding them. Furthermore, pilots will inform the ATS of this occurrence in order that other pilots are aware of a deteriorating situation. Although thunderstorms do not necessarily fit a macro scale, a phenomenon known as a 'line squall', which will be described, this takes on a much larger scale. This may start as an isolated (probably rather large) thunderstorm. West

Africa is an area where very marked line squalls can be encountered. The mechanism of this phenomenon is treated in greater detail in Chapter 15, 'Weather in Africa'.

The polar lows are described, and associated weather. Although individually they are a mesoscale phenomena, this term is now in general use for all vortex phenomenon in polar regions. The effect of temperate maritime climates in both hemispheres is described. The effects on general climate and seasonal variations are also described.

A schematic representation of the zonal wind system near the Earth's surface was described in Chapter 1, and can be seen in Figs 1.2 and 1.3. They comprise easterly trade winds in the tropics, calm subtropical high-pressure zones, mid-latitude westerlies and stormy low- and high-pressure zones close to the poles. In cross-section, these zones can be represented as three circulation systems covering the tropics, the mid-latitudes and the polar regions. (See Fig. 1.5). Incidentally, these general circulation features, both horizontal and vertical, are clearly recognisable on satellite images.

The weather in tropical regions is dominated by vertical circulation (the Hadley cell). The general motion consists of air rising near the equator to heights of up to 20 km ($\cong$ 65 000 ft) and then spreading out to 20–30° North and South latitudes before descending and flowing back towards the equator. The rising air at the equator is humid, and cools as it rises. This leads to the formation of towering shower clouds, which girdle the Earth and produce heavy rainfall in equatorial regions.

The precise position of this band of convective activity is known as the inter-tropical convergence zone (ITCZ). In some recent meteorological publications, the ITCZ is referred to as the inter-tropical confluence (ITC). It tends to follow the sun's latitude, although trailing about 4–6 weeks behind. It brings in rain to the dryer regions around 20° N and 20° S. Its movement is also linked to the distribution of the oceans and continents in the equatorial regions. (The ITCZ is described in greater detail in Chapter 5).

The descending air in the Hadley cell becomes dry and is warmed adiabatically, hence the regions around 20–30° N and 20–30° S latitudes are arid. Again, the desert regions stand out very clearly on satellite images.

The circulation in the tropics can be described in relatively simple terms because variations in surface temperature are relatively small. However, in extratropical regions global circulation of the atmosphere is controlled by different mechanisms. These are characterised by strong rotational motion which involves substantial temperature gradients. At low altitudes, this motion involves mobile areas of low pressure depressions and high pressure anticyclones. Above about 3 km altitude ($\cong$ 10 000 ft) it consists primarily of large waves (which will be described later) generally moving from west to east. These are superimposed upon a strong zonal current, the core of which is a 'jet stream' (see also Chapter 3).

The existence of long waves in the upper atmosphere has a fundamental influence on the weather patterns of the mid-latitudes, as they tend to guide the course of low-level eddies. The airflow patterns become very complex, reflecting both the changes of the seasons and the distribution of the oceans and continents. In particular, these factors exercise great influence in the northern hemisphere, because of the greater superficial land surface.

Looking at the jet stream again, it is observed to be strongest near the east coast of Asia and over the eastern United States and North Africa when the temperature gradients are greatest. In the summer, the jet stream in much weaker and is located further north.

The mean pressure patterns at sea level reflect very dramatically the influence of the oceans and continents. A further important consideration is the location and alignment of mountain ranges. This can be crucial, because mountain ranges can effectively block low level air mass migration. In winter, low pressure areas form over the relatively warm oceans and high pressure forms over the cold continental land masses, so the pattern is dominated by low pressure areas over the North Pacific and Iceland, and by high pressure (often very high pressure $\cong$ 1065 Hpa and higher) over Siberia. The high pressure area over Siberia has an influence on the whole of the northern hemisphere weather in winter, as will be described later. It is displaced far to the east, as there is no mountain barrier to Atlantic air moving in from the west, while the mountains to the south prevent any (appreciable) exchange with the Pacific and southern continents.

Over North America, the high nestles in the lee of the Rocky Mountains. In summer, the general pressure patterns are effectively reversed, with high pressure over the relatively cool oceans and lower pressures over the now warm continental land masses.

2.2 Atmospheric heat exchange processes

The exchange of heat between the surface of the Earth and the atmosphere occurs by radiation, conduction and convection. In the process of conduction heat passes from the warmer to the colder body *without the transfer of matter*.

Convection is a more important method of transfer of heat energy in the atmosphere. In this process *the body carrying the heat itself moves from place to place*. In the atmosphere pressure differences arise due to heating. With this process, warm air is forced to rise and cold air sinks to lower levels to replace it.

There are two types of heat: *sensible heat* which can be sensed or felt, and *latent heat*, which cannot be sensed directly. Latent heat or 'hidden heat' is the heat added to a substance when it changes its state from solid to liquid or from liquid to a vapour without changing its temperature.

Convection currents in the atmosphere not only transport sensible heat aloft; they also transfer upwards the latent heat stored in water vapour. This latent heat enters the atmosphere when water evaporates from the Earth's surface. Later, it is released in the upper air when water vapour condenses to form cloud.

The radiation from the sun provides the energy for the circulation of the atmosphere and the oceans. This is not lost, but has merely been changed into the form of heat energy or kinetic energy of the particles in motion. Solar energy may be transformed several times during heat exchange processes between the Earth and its atmosphere. Eventually, however, the solar energy absorbed by the Earth–atmosphere system is re-radiated back to space. By emitting the same amount as it receives, the system maintains its radiative balance.

This balance does not usually apply at individual latitudes. Between latitudes 0° and 35° in both hemispheres more energy is being absorbed than is being radiated out to space, and there is an energy surplus in these regions. Similarly, an energy deficit occurs in regions between latitude 35° and the poles.

The temperatures would develop if radiative equilibrium were achieved at each latitude without an exchange of heat would see a very large meridional temperature gradient (from equator to pole). However, the average meridional gradient, which is actually observed, is much lower than theory indicates. This is because heat is effectively transferred meridionally from low to high latitudes. The atmosphere and the oceans are involved in this energy transport. The meridional transfer of energy is assisted by the action of large-scale pressure systems (highs and lows), which develop in the regions where strong horizontal temperature gradients occur. Ocean currents also carry some energy away from tropical regions towards the poles.

The effect of radiation at the Earth's surface, and albedo

When solar radiation reaches the Earth's surface, it may be absorbed or totally reflected; this is dependent in the main on the nature of the surface. The *albedo* of a surface is defined as the ratio of the amount of global radiation from the sun and sky reflected by the surface to that which falls on it, i.e.

$$\text{albedo (of surface)} = \frac{\text{global radiation reflected by surface}}{\text{global radiation falling on surface}}$$

A great deal of the radiation falling on snow or ice is reflected. The albedo of snow surfaces ranges from over 0.80 for fresh snow to about 0.50 for old (dirty) snow surfaces. Other land surfaces, such as forests, grasslands, open fields and rocky deserts, have an albedo ranging from about 0.10 to 0.20. Dry sand values are in the range of 0.30, and a deep dark forest area about 0.05. The albedo of sea water depends on the angle of incidence of the incoming radiation, and varies from 0.05 when the sun is overhead to 0.70 when the radiation is near horizontal. The albedo of clouds is dependent on cloud type and thickness, but is deemed to range from 0.50 to 0.65.

Temperature differences between land and sea surfaces

The temperature increase of the Earth's surface when absorbing solar radiation is very variable. It depends in part on the distance to which heat penetrates and partly on the specific heat of the material. The specific heat of a substance is the quantity of heat required to raise the temperature of unit mass of the substance by 1°C. With the exception of hydrogen, water has the highest specific heat of any substance. It requires a relatively large amount of heat energy to raise the temperature of unit mass of water by 1°C.

Sand, depending on its colour, absorbs different amounts of radiation. Its specific heat is low and so its temperature rises rapidly when it is heated. In addition, it is a poor conductor and only a thin layer of sand absorbs the radiation. As a result, the temperature of a sand surface rises rapidly during the daytime. During the night-time, incoming solar radiation ceases and the sand loses heat by long-wave radiation. It therefore becomes increasingly colder during the night. Sand surfaces are therefore subject to large extremes of temperature between daytime and night-time. Similar effects occur when insolation falls on rock and other land surfaces.

Water absorbs a large proportion of the incoming radiation when the sun is at high altitude. However, it has a high specific heat and so its temperature only rises slowly. Some of the radiation penetrates the water to a depth of several metres, while mixing of the surface layers tends to spread any temperature changes through a considerable depth. In addition, some of the heat energy gained by the water is converted into latent heat during the evaporation process.

The temperature of the sea therefore does not rise as rapidly as that of a land surface during the daytime. At night, the incoming radiation ceases and heat is lost by radiation. However, there is usually a large store of heat energy below the water surface, and so very little change occurs in the surface temperature. The variation in the temperature of sea surfaces on a diurnal scale is therefore very small.

2.3 Long waves (Rossby waves)

On occasions, there may be massive poleward advection of air in some parts of the hemisphere, but compensating flow towards equatorial regions in other areas. Large-scale air movements are associated with long waves or Rossby waves which girdle each hemisphere. They can be found in the mid- and high troposphere in mid-latitudes. Laboratory experiments confirm that such waves are an inherent feature of any rotating fluidic medium that is subjected to a thermal gradient, i.e. a rotating dishpan heated on its outer wall and cooled at the centre can be made to produce either symmetric or standing wave patterns *depending on the rate of rotation and the temperature difference across the fluid.* These waves can be shown to be responsible for the main heat transfer between the warm rim and the cool centre.

On a global scale, these wave patterns are subjected to the following situations. Air moving towards the pole is deflected to the right by the Coriolis force. As air is displaced away from the equator, the curvature becomes anticyclonic (Fig. 2.1). Air moving towards the equator finds the Coriolis force decreasing, therefore the deflection decreases and the curvature becomes cyclonic, now moving the air polewards. In this manner, large-scale flow tends to oscillate in a wave motion.

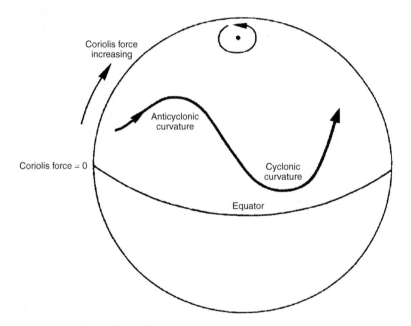

Fig. 2.1 Schematic of the formation of long waves in the tropospheric westerlies.

The wavelength spacing is dependent on latitude and the speed of the zonal current. The speed of the wave (or phase speed) is dependent of the size of the wave. Waves normally move from west to east; however, very long waves are slow moving, and can in fact move in retrograde (east to west). As an example, at 45° North latitude, for a wave with a zonal velocity of 9 kt the wavelength would be 3120 km. If the zonal velocity increases to 24 kt, then the wavelength would increase to 5400 km.

The Earth's circumference limits the circumpolar westerly flow to between three and six major Rossby waves. The importance of these waves is that they effect the formation and movement of surface depressions. It must also be mentioned that the huge Siberian anticyclone in winter causes a major disruption in long waves.

In response to the influence on the atmospheric circulation of large mountain barriers, such as the Rocky Mountains and the Tibetan plateau, the main *stationary* waves tend to be located about 70°W and 150°E. The waves are also sensitive to large-scale heat sources, such as warm ocean currents (in winter) or landmasses (in summer).

Another point to bear in mind is that (in the northern hemisphere) on the eastern limb of troughs in the upper westerlies, the flow is normally divergent, since the gradient wind is lower than pure geostrophic around a trough, but higher than pure geostrophic around a high (or ridge). This has the effect that the sector ahead of an upper trough is a favourable location for a surface depression to form, or deepen, and it will be noted that the mean upper troughs are significantly positioned just west of the Atlantic and Pacific polar front zones in winter.

Weather patterns in the mid-latitude zone (35–55°N) can usually be characterised by seasonal averages (the definition of climatology); however, at any one time the patterns can show an enormous variety. Within a diurnal time period, it is possible to show that these exhibit irregular quasicyclic changes, which can best be measured in terms of *pressure differences* across the zone. With a large pressure difference, there are strong sea-level westerlies and a corresponding long wavelength pattern in the upper atmosphere.

With a small pressure difference, there is a breakdown of the sea-level westerlies in closed cellular-like patterns and a corresponding meandering shorter wavelength aloft. (See Figs 2.2a, b and c). It has been observed that the breakdown of the westerlies leads to prolonged periods of abnormal weather with compensating extremes around the globe. It can be seen that to observe long waves would be important in the longer-range forecasting as long waves are much slower to change character, whereas surface weather phenomena can change rapidly in response to diurnal variations for one reason or another.

Referring to Figs 2.2a, b and c, the long wave (cellular patterns) develops from '2.2a' through '2.2d' in about 5 to 8 weeks, and is especially active in February and March in the northern hemisphere.

The nature of the changes shows zonal westerlies over middle latitudes, (2.2a) and this is described as having a high zonal index. Waves develop forming troughs and ridges, which become accentuated, ultimately splitting up into cellular patterns with pronounced meridional flow at certain longitudes (Fig. 2.2d). This situation is described as having a low zonal index.

In '2.2a' the polar jet stream lies north of its mean position, as do the westerlies. The westerlies are strong, and the pressure systems are orientated east–west, so there is little air mass exchange through north–south.

In '2.2b' and '2.2c' the flow pattern oscillates with deeper excursions. The jet stream expands and increases in speed. In '2.2d', there is a complete break-up of the zonal westerlies, and the pattern fragments into discrete cells. The significance of the evolution of this upper air pattern can be seen on the surface. At high latitudes there would be deep blocking warm anticyclones and at lower latitudes quasi-stationary deep occluding cold depressions. Further upper air characteristics are described in Chapter 3.

From observations, and also theoretical studies, a result of the variation of the Coriolis force with latitude sees cyclones entrained in the westerlies having a tendency to move towards the nearer pole, and anticyclonic cells move towards the Equator. This then can explain how the subtropical anticyclonic cells are regenerated. Furthermore, in the troposphere, there is a relationship between the position of the subtropical high-pressure cells and the meridional temperature gradient. When the gradient is strong, there is an equatorward movement of high pressure, and when it is weak a reverse tendency.

The seasonal changes in heat sources can be seen in the surface cellular patterns. They exhibit elongation along the north–south axis, and highs tend to remain stationary in the northern hemisphere in summer over oceanic areas. The land masses are then warm and exhibit low pressure. The meridional temperature gradient becomes weak.

In the winter, the meridional temperature gradient is much greater, and the zonal flow is therefore much stronger because the landmasses are colder. This produces an east–west orientation of the pressure cells. This pattern is not recognised to the same degree in the southern hemisphere because of the lower superficial landmass.

There is one additional phenomenon in the general air circulation in the northern hemisphere which must be given special attention, and this is the monsoon that affects the Indian subcontinent.

The essential reason that the south-west monsoon (summer monsoon) is so much stronger over India than elsewhere in the tropical latitudes is because of the deployment of adjacent landmasses, and particularly the mountain ranges. To the north of India lie the Himalayas, with the Tibetan plateau further north. As Asia heats up, and the thermal gradients (upper atmosphere) increase, the jet stream switches from south of the Himalayas to the north, and progresses towards northern Tibet. The moist tropical air and associated heavy rains along the ITCZ also move north, in fact move further north over India than anywhere else in the world. The Indian monsoon will be discussed in greater detail in Chapter 17, 'Weather – Arabian Gulf to Singapore'.

The circulation in the southern hemisphere can be described more easily, for several reasons. In the first instance, because there is less superficial land area, it has the effect of reducing the contrasts of oceans and continents, particularly in mid-latitudes, and secondly, the Antarctic is virtually symmetric about the South Pole, which has the effect of reinforcing the zonal circulation of the atmosphere. Therefore, the air flow patterns both at sea level and aloft are more uniform. The westerlies are, however, stronger and show less variation between summer and winter. Furthermore, pressure differences across the mid-latitude zone are less marked, and there is less of a

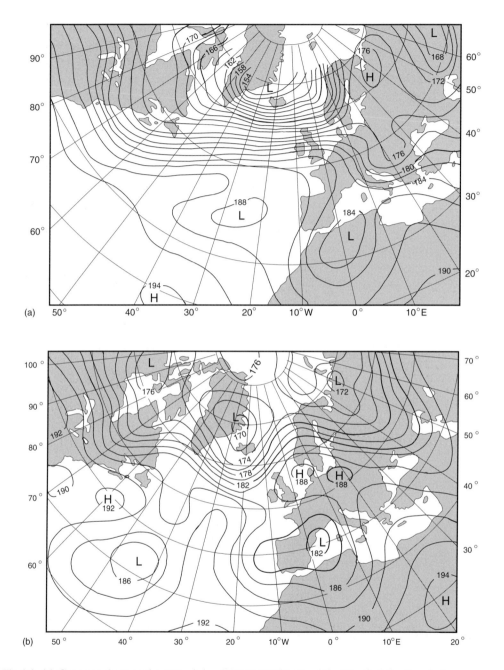

Fig 2.2 (a) Contour chart at the 500 mb level ($\cong$ 18 000 ft). Over the North Atlantic winds are essentially westerly – high zonal index. Jet stream situated north of its mean position. (b) Contour chart at the 500 mb level ($\cong$ 18 000 ft). Over the North Atlantic the flow pattern begins to oscillate with a southerly excursion. The jet stream expands and accelerates.

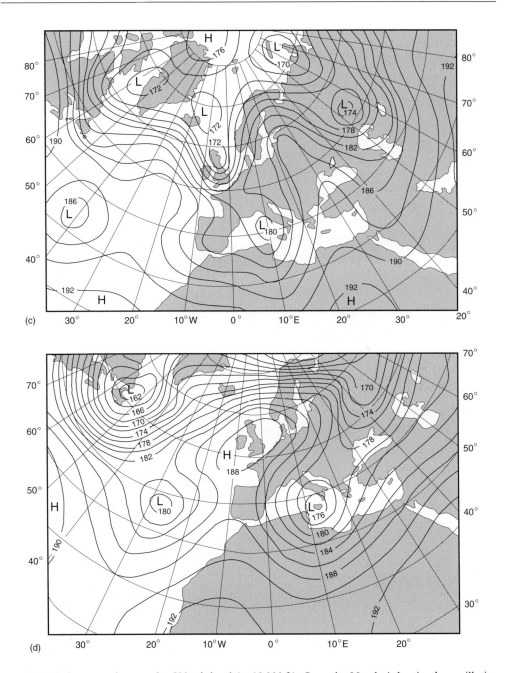

Fig. 2.2 (c) Contour chart at the 500 mb level ($\cong$ 18 000 ft). Over the North Atlantic, the oscillation continues, and the jet stream increases in speed. (d) Contour chart at the 500 mb level ($\cong$ 18 000 ft). Over the North Atlantic, the flow patterns now fragment into discreet cells. Warm blocking anticyclones in the north and deep cold depressions at lower latitudes.

tendency for the zonal flow to break down into meandering patterns that result in abnormal weather.

2.4 Ocean currents

One aspect, which has dramatic effect on climate, is ocean currents. We must remember that oceanography is a science in its own right, but there are some special areas that should be mentioned. One is the confluence of two main ocean currents, the warm Gulf Stream and the cold Labrador Current in the area to the north and west of Newfoundland. This is the area of the New-foundland fishing banks. Warm moist air flowing from over the Gulf Stream when passing over a very much colder current forms advection fog. This is a classic example of an advection process. It can form any time of the year, and at any time day or night.

The other points are the forces that determine ocean currents. There are three main forces: one is the prevailing wind, the second is the Earth's rotation, and the third is the differences in sea density. Winds drive immense bodies of water before them forming surface currents. At the same time the Earth's rotation deflects moving bodies of water to the right in the northern hemisphere and to the left in the southern hemisphere. We need not explore sea density as this is in the realm of oceanography.

Ocean currents are usually called 'drifts'. The effect of the trade winds, which force sea water to flow along in the general direction of the wind, is a continuing process. However, the wind carries on over the landmasses, and the sea is not only deflected by the Earth's rotation, but is also forced to deflect along the coastline of a major landmass. A large-scale circulation is set up whereby the *warm sea affects western coasts, and cold seas affect eastern coasts* (see Fig. 2.3). The coasts affected are subjected to the cold or warm adjacent ocean and climate is so modified. The general features of these modifications are as follows.

Cold water coasts
Over land, due to the low rate of evaporation from the nearby relatively cold ocean, the air in these areas has a low water vapour content. As a result, there is little or no cloud or precipitation, and deserts can extend right up to the coasts. Night-time cooling may be sufficient to produce low stratus by dawn but it soon disperses after sunrise. Over the sea, the air that has moved over the cold current will have experienced appreciable surface cooling and widespread advection fog (or low stratus and drizzle) often occurs both by day and night. This fog may drift over the adjoining land when the land cools at night.

Warm water coasts
Over both land and sea, humidities are high due to the rapid evaporation from

Fig. 2.3 Ocean current circulation.

the relatively warm ocean. Over land by day and over the sea, surface temperatures are relatively high and cumuliform cloud with thunderstorms, squally winds, heavy precipitation etc., results. Over land at night, the diurnal variation of surface temperature can cause clouds to collapse into stratiform cloud which will then redevelop to cumuliform cloud again the following day. Over sea at night the cumuliform cloud tends to persist.

From Fig. 2.3, some of the important ocean currents can be identified. 'A' is a cold water current, the Oya Shio current, which affects the coasts of Japan; 'B' in the opposite direction is warm water, the Kuro Shio current. 'C' indicates the cold Labrador current, and 'D' the warm Gulf Stream. These currents provide the mechanism to produce the classic advective fogs in these areas. 'E' is the cold Falklands current penetrating the mainly warm opposing current; 'F' is the Antarctic circumpolar, or West Wind drift.

2.5 Arid climates

The most marked and important arid zones of the world are situated around latitudes 30°N and 30°S, where large areas are dominated by dynamic anticyclonic subsidence, but it must not be assumed that arid areas cannot occur elsewhere. Aridity is a normal phenomenon; it is atmospheric precipitation which needs a special explanation.

The arid zones in the subtropics are very well marked and observed. It is interesting that these zones extend over the subtropical oceans as well as over the subtropical landmasses. This indicates that an absence of surface water is not the main cause of an absence of rainfall, for indeed the actual cause of general aridity in the subtropics is the widespread subsidence of the atmosphere in the descending limbs of the Hadley cells. The subsidence effectively inhibits extensive low-level convergence, and it is therefore almost impossible for extensive cloud systems to form with sufficient depth and consequent rainfall.

If the surface consists of the open ocean, then evaporation will be high because of the high net radiation; the surface layer will become moist and this moisture will be available for use by occasional synoptic disturbances. In contrast, if the underlying surface is completely arid, there will be no local evaporation and no moisture gain by the surface layers of the atmosphere.

It is therefore possible to distinguish two general types of arid climate:

(1)   A maritime arid climate.
(2)   A continental arid climate.

The maritime anticyclones are the source regions of the trade winds, and the evaporated water is carried towards the Equator. The moisture in the layer below the anticyclonic inversion can be utilised by any synoptic disturbances which may develop. On the other hand, over large arid continents all the net radiation is available for heating the soil and the air in contact, and as a result the surface temperatures becomes high with low humidity. If surface convergence does occur over an arid landmass, there is probably not enough water vapour in the air to form cloud and rainfall; moisture must be imported from outside the desert areas if rain is to be formed. Conditions are therefore more extreme in large deserts than over the oceans.

2.6 Deserts

There are no desert areas that are completely rainless, although in extreme cases several (or many) years may lapse between individual rainstorms. The subtropical anticyclones do not have great seasonal variations, so the desert mid-zones are liable to remain almost rainless. However, rainfall is apt to increase and become more seasonal towards the poleward and equatorial limits of the deserts.

Rainfall within desert regions normally results from disturbances arising from outside the true desert areas. As an example, troughs from the middle latitudes or upper cold pools are the mechanisms providing rainfall to the poleward margins of the deserts. Synoptic disturbances forming near the Equator can bring rainfall to the equatorward margins. The enormous energy in tropical storms can result in rain reaching areas that may have been truly arid for years, i.e. the western desert region of Australia (through tropical cyclones that have formed in the southern Indian ocean and re-curved to the east). See Chapter 6, 'Tropical Storms'.

Middle latitude troughs are most intense and nearest to the Equator in winter, therefore the poleward margins of the deserts usually experience a winter rainfall maximum. The equatorial trough tends to move north/south with the sun; in contrast to the poleward margins, the equatorial limits of the desert areas normally experience summer rainfall maximums. This results in two distinct rainfall maxima in desert regions.

On the equatorward side of the subtropical high-pressure areas, there are generally easterly winds, and where these originate over landmasses they are extremely dry. This dryness is carried over the oceans to leeward of the landmasses. The trade winds pick up moisture by local evaporation from the ocean surface as they travel westward and because of this, deep surface moist layers are most usual in the western parts of the tropical oceans.

Tropical disturbances depend partly on the release of latent heat for their generation and maintenance. They tend to become more frequent towards the western ends of the subtropical oceans because deep moist layers exist. The western edges of the oceans therefore receive large amounts of rainfall from disturbances and form a break in the normal subtropical arid belt.

2.7 'Trade winds' and 'trade-wind inversions'

The tropical maritime air such as found over the vast oceanic regions does not normally possess a deep moist layer, because usually there is an inversion about 2 to 3 km above the surface. Below the inversion, moist air is trapped; above the inversion, the atmosphere is dry and cloudless. As this phenomenon is found typically in trade-wind areas, it is known as the trade-wind inversion. The trade winds occur on the equatorial sides of the subtropical anticyclones, and in the greater part of the tropics (Fig. 2.4).

In general, the trade winds blow from ENE in the northern hemisphere and from ESE in the southern hemisphere. Trade winds are noted for their extreme continuity in both speed and direction; there is no other climatological wind flow system on Earth where this extreme continuity exists. This is the result of the permanence of the subtropical anticyclones; interruptions do occur but are the result of a major atmospheric disturbance.

Winter is when the subtropical anticyclones are most intense, so therefore the trades are strongest in winter and weaker in summer. The trade-wind inversion is now known to be of great importance in the meteorology of the tropics. See Fig. 2.5 for the vertical structure of the trade-wind air. Broad-scale subsidence in the subtropical anticyclones is the main cause of the very dry air above the trade-wind inversion.

The subsiding air normally meets a surface stream of relatively cool maritime air flowing towards the Equator. The inversion forms at the meeting point of these two air streams, both of which flow in the same direction, and the height of the inversion base is a measure of the depth to which the upper current has been able to penetrate downwards. The main source regions of trade winds and the trade-wind inversion are the eastern ends of the sub-tropical anticyclonic cells, in particular along the western edges of the Americas and Africa. Upper air streams flowing towards the Equator tend to subside and diverge. Referring to Fig. 2.6, one can see the height of the base of the trade-wind inversion over the Atlantic (in feet). It also shows that the height of the trade-wind inversion increases towards the Equator.

The trade-wind air streams tend to converge in the equatorial trough (low pressure region) over the oceans, particularly noted in the Central Pacific, where the convergence of these air streams is most marked, and in this area the term inter-tropical convergence zone (ITCZ) is applicable. Remember that the ITCZ may be called the inter-tropical confluence (ITC) in some meteorological texts.

Elsewhere the convergence seems to be discontinuous. Equatorwards of the main trade-wind zones are the regions of light, variable winds, known traditionally as the doldrums, and much feared in past centuries by the crews of sailing ships (see Fig. 2.4). The seasonal extent of the doldrums varies considerably; from July to September they spread westwards into the Central Pacific while in the Atlantic they extend to the coast of Brazil. A third major

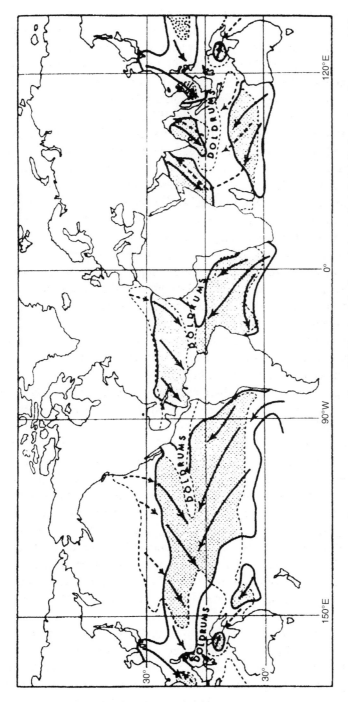

Fig. 2.4 Location of the trade-wind belts and doldrums. The solid (January) limit, and the dashed (July) limit enclose the areas within which 50% of all winds are from the predominant quadrant. The stippled area is affected by the trades in both months. Streamlines are shown by dashed (July) and solid (January, or both months) arrows. (From Crowe, 1949 and 1950.)

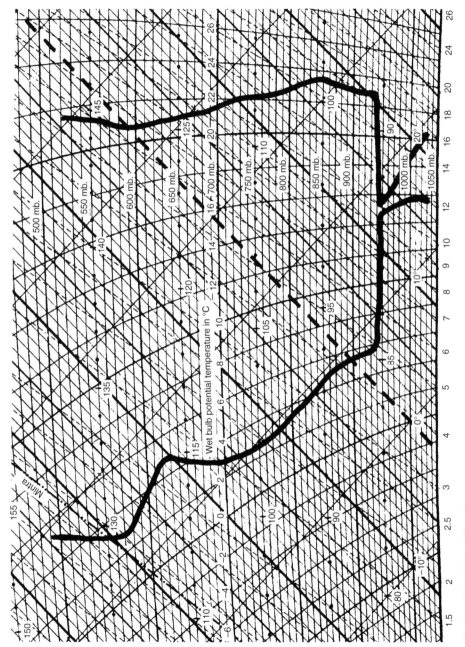

Fig. 2.5 Vertical profile of temperature and dew-point temperature through a trade-wind inversion, showing the very marked characteristic inversion at about 950 mb. Below the inversion the air is moist. Cloud will be contained below the inversion, dry (descending) air above the inversion. (After Riehl, 1954.)

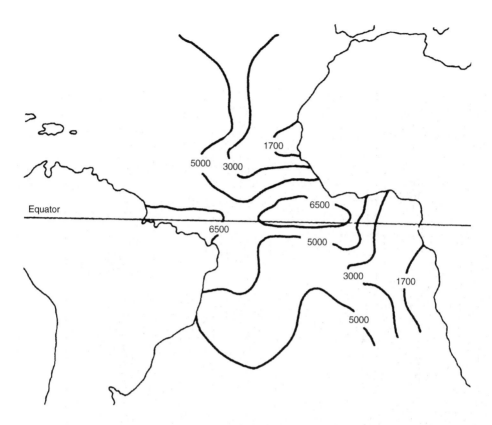

Fig. 2.6 Heights of the trade-wind inversion (approximate heights in feet) over the tropical Atlantic Ocean. The trade-wind inversion increases in height towards the equator. The inversion will break down in the event of an active ITCZ, easterly waves, or tropical cyclones. (After Riehl, 1954).

doldrum zone is located in the Indian Ocean and Western Pacific. In March to April it stretches 16 000 km from East Africa to 180° longitude and it is again very extensive during October to December.

The trade-wind inversion at low level can have unfortunate effects on cities near its source regions. One classic example in summer is when a low-level inversion spreads over Los Angeles and very effectively stops any vertical exchange of air from taking place. The combustion products from industry and domestic households, plus the exhaust fumes from two million motor cars rise into the atmosphere below the inversion and remain trapped within it. Since the surface layer is already naturally hazy, the result is the formation under strong tropical sunlight of the notorious Los Angeles 'smog' (a word which is derived from a combination of the words 'smoke' and 'fog'). Other tropical cities can suffer from similar smogs due to the trade-wind inversion.

High mountains on tropical islands penetrate through the trade-wind inversion into the dry air above, and as a result desert or semi-desert conditions are found on the mountain tops, with a zone of high rainfall on the lower mountain slopes below the inversion. Hawaii is a good example; in the centre of the island at 3000 m the annual rainfall is about 250 mm, whereas below the inversion at 1000 m the annual rainfall is 7500 mm or more.

2.8 Thunderstorms

At any given moment, there are about 2000 active thunderstorms around the world. They are a daily occurrence in the tropical latitudes, and seasonal in temperate latitudes. The primary requirement for the development of a thunderstorm is a *conditionally unstable lapse rate*. The atmosphere must have a high vapour content, at least in the lower levels. Also, a trigger action to lift the air, i.e. solar heating, orographic displacement, convergence or fronto-genesis.

Under these conditions, an air parcel lifted to condensation at low levels is capable of continuing its ascent through several kilometres, its buoyancy being maintained by latent heat release as vapour condenses. Normally an air parcel that is lifted would cool due to expansion as the surrounding pressure drops with height (adiabatic cooling). However, latent heat release is enough to give the rising air a positive buoyancy with respect to the ambient air, and it will continue upwards, and even accelerate. The temperature of rising air can be from 2° to 12°C higher than that of the ambient air at about 5 km altitude. As the air parcel continues to rise, the actual water vapour content (AWVC) decreases, but the rising air remains saturated. As water vapour decreases with altitude, so does the available latent heat, and the rising parcel of air becomes less buoyant, but inertia can raise the parcel to heights where negative buoyancy would occur and beyond. This can be seen in tropical latitudes where fully developed thunderstorms have 'domed' tops where rising air has pushed into the stable stratospheric layer and produces a smooth lifted cloud cap (*pileus*).

Severe storms can result from particular situations. One example is as a result of *eroding a capping inversion*. This process is one in which there is a low-level stable layer that initially prevents the moist boundary layer from penetrating to an unstable layer above. An example of this mechanism can be found in a returning polar maritime air mass (North Atlantic). The air mass is unstable due to heating from a progressively warmer sea as it tracks south, then the returning section of its track is heading in a northerly direction, after sweeping around a low pressure area to the west of Ireland. It is now being cooled from below making the lower levels stable, but instability still exists above. Extremely unstable lapse rates can develop above such inversions. If the stable layer is penetrated by an updraught, the parcel of air will accelerate upwards.

Another situation can be where *frontal convergence* takes place. This is where a colder air mass meets a warm, moist air mass. The warm, moist air mass is forced to rise over the advancing cold (more dense) air mass making the former stable warm air become unstable. During the day, heating of a land surface will also enhance lifting. As there is also cooling of the upper levels taking place this too enhances instability. Lifting can also cause mid-level conditional instability, and this can be seen where *altocumulus castellanus* is forming.

When air is warmed from below during daytime over land, and further cooling takes place at high levels, due to overriding cooler air, the lapse rate becomes increasingly unstable. Over the sea, an air mass can become unstable as it tracks over warmer ocean currents. This is a common occurrence in a cold outbreak of air tracking over warmer water found in Atlantic cold fronts forming shower clouds.

Along the west coast of Africa, warm moist air along the leading edge of the ITCZ when tracking northwards, penetrates inland. The air ahead is dry and exhibits a stable lapse rate; however, the maritime air is now being heated from below forming cumulus in the early stages, but as the moist air penetrates further inland, its depth also increases. When this depth exceeds $\cong 10\,000\,\text{ft}$, towering cumulus and cumulonimbus clouds are formed. Sufficient latent heat is being released to produce buoyancy up through the maritime air, and also though the dry (and cold) continental air aloft, where the updraught will accelerate.

A similar mechanism can be found in the southern USA. Air originating over the high plains in the Midwest, which is essentially stable, is undercut by warm, moist air from the Gulf of Mexico. The resulting lapse rate becomes very unstable, and produces huge thunderstorms that can lead to 'super cells', and the infamous tornadoes.

A further ingredient to create severe storms is *vertical shear of the horizontal wind speed*. Thunderstorms can exist without vertical wind shear, and in fact it was thought at one time that vertical shear inhibited the full development of a thunderstorm. To a degree this is true but it is the shear *in direction* not speed. When a storm is sufficiently developed and glaciation takes place (at about $-20°C$ level) ice is being formed by direct evolution from the vapour to solid crystals (sublimation). From this zone upwards, more of the precipitant consists of ice rather than water, the solid ice particles acting now as a freezing nucleus, and more and more of the water drops are becoming frozen into graupel and hail.

This is the stage where vertical shear of the horizontal wind becomes more important. The developing thunderstorm column is now tilted, or leaning towards the downwind wind vectors, and this results in rain falling at an angle to the vertical profile of the storm. Rain is therefore falling into unsaturated air ahead and evaporates. This segment of air is now being cooled and together with the falling rain causes a downdraught. How far this downdraught extends and its speed are dependent (mainly) on the relative humidity. If the relative humidity is high, the evaporative cooling is less; if it is relatively dry, the evaporative cooling is greater, and the descending air can reach the surface and be some degrees cooler than the surface environment and spill outwards along the ground. This is the *gust front*.

If there is no (or very little) vertical shear of horizontal wind speed, then the falling rain is descending into saturated air, and a substantial downdraught will not be formed. This produces a regular shower cloud.

Severe storms of long duration *must have marked shear*. The downdraught must cool the air ahead of the storm, and as this is essentially dry the evaporative cooling enhances the downdraught. The updraught provides the rain for the downdraught. If the gust front is positioned such that it becomes enhanced by the downdraught, then we can see a sequence of events that will provide a positive feedback mechanism to sustain the storm. Colder outflows from the storm (the gust front) move out from the storm column and undercut the ambient warmer air, forcing it to rise (convergence) and thus create further storm cells.

Individual storm cells are short lived, but in this mechanism where new cells are continually being created the squall line is sustained and can last for a considerable time. Once this feedback mechanism is set in motion, squall lines can travel great distances, over 1000 km. The West African line squalls are of this type. See Chapter 15, 'Weather in Africa', Section 15.6 on West African line squalls, p. 145.

To predict the general direction of storms, it is generally assumed to follow the wind vector at 700 mb ($\cong 10\,000$ ft); however, with multicell squall lines this is not the case. The movement of the squall line is determined by the direction of the gust front. It is the gust front that provides the lifting mechanics to form new cells, so not only is this mechanism capable of forming a line of storms, but the line as a whole will advance in the direction of the outflowing gust front. This can be some 40° or 50° removed from the 700 mb vector (Thorp & Miller, 1978).

2.9 Polar lows

Polar lows were originally thought to be of a convective nature, but Rasmussen (1979) suggested they are initiated by deep cumulus convection through conditional instability of the second kind (CISK), and indicated some similarities between polar lows and tropical cyclones. CISK is a positive feedback mechanism, which develops between cumulus-scale convection and the large-scale circulation. It is important in the development of tropical cyclones as well as polar lows. See also to Charney and Eliassen (1964).

There is in fact a whole spectrum of mesoscale lows, which this term can be applied to. Nevertheless, some general characteristics of polar lows can be identified.

(1) Polar lows form poleward of the main polar front.

(2) They have a horizontal scale of 500–1000 km and sometimes even less.

(3) They form over the ocean and decline rapidly after making landfall.

(4) They usually have surface winds in excess of gale force, along with heavy snow.

(5) In the northern hemisphere they are exclusively a winter season phenomenon, but they seem to occur throughout the year in the southern hemisphere.

(6) Polar lows are non-frontal.

Since the 1960s the term polar low has been used to describe an increasingly broad range of phenomena and it is now used to describe virtually all mesoscale vortices in polar regions.

Where do they form?

Polar lows occur generally in certain preferred ocean areas of the polar regions. However, they have traditionally been associated with the North Atlantic and in particular the Barents and Norwegian Seas and the area south of Iceland, since this is where they were first observed, and where the most extreme examples are found. A synoptic chart showing a polar low off the coast of Norway is shown in Fig. 2.7, from which the characteristic 'bullseye'

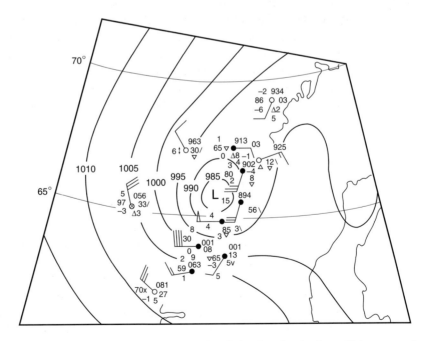

Fig. 2.7 Synoptic chart showing a typical well-developed polar low off the coast of Norway. (From Rasmussen 1979)

appearance of the low can be seen, as there are sufficient observations available to establish the centre accurately. A well marked and symmetrical polar
low viewed by NOAA-9 satellite can be seen in Fig. 2.8.

Fig. 2.8 NOAA-9 Infra Red satellite image taken at 0418 UTC on 27 February 1987. This particular
polar low shows a remarkable symmetry. It is entrained in a cold air mass and positioned just off
North Cape, North Norway. *Image © of the University of Dundee, and reproduced with permission.*

Why do they form?

Formation of polar lows occurs when cold air flows over relatively warm
water. Off the north coast of Norway, air temperatures as low as $-30°C$ from
ice-covered areas can be advected over the water of the warm Gulf Stream
which has a surface temperature approaching $10°C$ in this region. This gives
rise to strong low-level baroclinic zones (horizontal thermal gradients) and
intense surface fluxes, both of which are important requirements in the
development of polar lows. From satellite imagery, it has been noted that

many small-scale vortices form on the baroclinic zones near the ice edge and some develop into polar lows after tracking some distance over open water. This is regarded as the classic mechanism to form polar lows in the Barents Sea and Norwegian Sea (Rasmussen, 1985). The polar low will exhibit extensive deep cumulonimbus clouds and associated heavy precipitation with hail.

The very vigorous convective systems occurring in the northern hemisphere are unlikely to form in the southern hemisphere because of the more zonal ocean currents. This is a limiting factor in the poleward movement of warm tropical water, thus giving a much smaller atmosphere/ocean temperature difference. Baroclinic systems do however form, and are often seen in the Weddell Sea (Turner *et al.*, 1991).

2.10 Temperate maritime climates

Some general notes about climates of the temperate oceans with those of the neighbouring land areas are given below. The North Atlantic, the North Pacific and the Southern Ocean around Antarctica have a substantial influence over the particular neighbouring land areas.

Because of the generally westerly flow of the atmosphere, maritime influences are most marked on the eastern sides of the oceans, and are particularly strong in Western Europe, western North America, the southern parts of western South America, New Zealand, and the extreme south of Australia. High mountain ranges limit oceanic influences to the coast in North and South America, and the other land areas in the southern hemisphere are extremely limited, so the only extensive land area with a temperate maritime climate is found in Western Europe.

General climate

These climates are characterised by oceanic or suboceanic temperature regimes with mild to moderately cold winters and moderately warm to warm summers. The more oceanic climates have an autumn and winter maximum of precipitation, the less oceanic climates have a summer to autumn maximum. These climates are not generally given to extremes of heat and cold. The climatic type is both well developed and well documented over the North Atlantic and Western Europe and can be taken as typical of the other areas.

As most of Europe is open to invasions of maritime zonal westerlies, the change eastward from marine to continental climates is very gradual, and marine effects are still obvious in Eastern Europe. It is therefore difficult to fix an eastern boundary for this climatic type but it can be assumed to be at about longitude 16°E, which is quite a long way into Europe. East of this longitude, the climate is dominated by continental air masses. The northern limit in Europe can be taken as the Arctic Circle, and the southern limit can conveniently be taken along a line through the Alps and the Mediterranean Sea.

Seasonal variation
Temperate maritime climates show marked seasonal variations, and seasons can change with marked abruptness. These sudden changes are present in both temperature and rainfall records. Some appear to be completely random, while others occur at nearly the same date each year. It is most likely that the upper flow patterns in middle latitudes have an irregular cyclic rhythm, and that this is reflected in surface pressure distribution and weather. See section 2.3 on long waves.

References

Charney, J.G. & Eliassen, A. (1964) On the growth of the hurricane depression. *Journal of Atmospheric Science*, **21**, 68.

Crowe, P.R. (1949) The trade wind circulation of the world. *Transactions & Papers, Institute of British Geographers* **15**, 37–56.

Crowe, P.R. (1950) The seasonal variations in the strength of the trades. *Transactions & Papers, Institute of British Geographers* **16**, 23–47.

Rasmussen, E.A. (1979) The polar low as an extratropical CISK disturbance. *Quarterly Journal of the Royal Meteorological Society* **105**, 531–549.

Rasmussen, E.A. (1985) A case study of a polar low development over the Barents Sea. *Tellus*, **37A**, 407–418.

Riehl, H. (1954) *Tropical Meteorology*. McGraw-Hill, New York (393 pp.)

Thorpe, A.J. & Miller, M.J. (1978) Numerical simulations showing the role of down-draught in cumulonimbus, motion and splitting. *Quarterly Journal of the Royal Meteorological Society* **104**, 837–893.

Turner, J., Lachlan-Cope, T. & Rasmussen, E.A. (1991) Polar lows. *Weather* **46**(4), 107–114.

Chapter 3

Upper Winds and Jet Streams

Further points about atmospheric physics are discussed in this chapter to explain the theory behind *thermal winds*, their alignment and the contribution to the real winds aloft. The concept of '*thickness*' is discussed, and example charts are shown to indicate the correlation with surface synoptic features. This leads to a discussion on the location of *jet streams*, the high level, high-speed ribbon of air flowing under just below the tropopause.

3.1 Upper winds

It is a common observation in mid-latitudes that clouds at different levels move in quite different directions. The wind speeds at these different levels would also be quite different.

The gradient of wind speed with height is referred to as a *vertical windshear*. However, vertical windshear can also be a change in direction, or in both speed and direction. The amount of shear is a function of the mean temperature throughout the layer or column of air being considered.

As height is increased, the temperature decreases. A component of the upper wind is the thermal wind, which is deemed to blow in the northern hemisphere with the area of lower temperature on the left (opposite in the southern hemisphere). Considering the northern hemisphere, where there is colder air towards the pole, then a westerly surface (geostrophic) wind will increase with height, because the thermal wind component is increasing in the same direction. If the surface wind is easterly, then the geostrophic wind will decrease with height, because the thermal component is in opposition. The wind could decrease to zero and then begin to increase as a westerly wind.

To take this a stage further, if there is low pressure at the surface with a cold column of air, then the wind which is blowing anti-clockwise around the low pressure area is being enhanced by the thermal component, which is also blowing anti-clockwise around the area of low temperature. As the temperature decreases with height, then the thermal wind increases, thus enhancing the geostrophic wind with height.

If there is high pressure at the surface with a cold column of air, then the wind which is blowing clockwise round the high-pressure area is in opposition to the thermal component, which is blowing anti-clockwise around the area of low temperature. As height is increased, temperature decreases, the thermal component will increase and being in opposition the geostrophic wind will first of all decrease and then reverse with height.

There is a relationship between the mean temperature of a column of air and surface and tropospheric pressure systems, for example:

(1) If there were high pressure at the surface with a cold column of air, then the high pressure weakens and becomes a low-pressure zone aloft.
(2) If there were a low-pressure area at the surface with a cold column of air, then the low pressure will intensify with height.
(3) If there were high pressure at the surface in a warm column of air, then the high pressure will intensify with height.
(4) If there were low pressure at the surface with a warm column of air, then the low pressure weakens aloft and becomes a high-pressure zone.

In the case of the very cold anticyclone over Siberia in winter, high pressure is present at the surface, but the high pressure weakens aloft and becomes a low-pressure zone in the cold column of air. The upper wind direction would be *westerly* on the southern flanks of the upper low-pressure zone (see Fig. 3.10). In fact, these winds can be very strong and form a jet, which divides to either side of the Tibetan Plateau. (Jet streams will be described later in the chapter.)

During the northern summer, an area of low pressure replaces the Siberian anticyclone, as the land is becoming warm. With surface low pressure in warm air, the low pressure weakens aloft and becomes a high-pressure zone. The winds aloft on the southern flanks of the upper high-pressure zone become *easterly*. The northern excursion of upper air westerlies in winter in this region is $\cong 35°$N. However, the northern limit of the easterlies in summer in this region is $\cong 30°$N. The centre of high pressure in the winter in this region is further north than the centre of the low pressure in the summer.

Earlier, in this chapter, it was stated that the amount of shear is a function of the mean temperature in a column of air. This important relationship is now explained with upper air 'contour charts' and 'thickness charts'.

Contour charts

Until the advent of regular upper air soundings, meteorological forecasting was based on the knowledge observed from the ground and forecasts were merely an extrapolation of these conditions. The development of the radio-sonde which could be tracked accurately by radar and the demand for accurate upper air forecasts has meant that much more precise attention can now be given to conditions at varying heights above the ground as well as to surface conditions.

On a surface chart showing isobars (lines joining points of equal barometric pressure reduced to mean sea level in ambient atmospheric conditions), then as the spacing of isobars indicates the pressure gradient, it follows that a suitable scale (geostrophic scale) could be applied and the windspeed extracted at that point. Wind direction is the local alignment of the isobars. However, the construction of isobaric charts at fixed heights above the Earth's surface and the measurement of upper winds by means of a geostrophic scale as on surface charts is not as convenient at upper levels due to the variation of the density of the air with height. The geostrophic wind relationship is:

$$V_g = \frac{1}{2\Omega\rho \, \sin\phi} \times P_g$$

where V_g = geostrophic wind speed; Ω = Earth's rotation in radians p/s; ρ = air density; ϕ = latitude; P_g = pressure gradient.

Thus we can only say that the geostrophic wind is proportional to the pressure gradient (or inversely proportional to the distance between the isobars) when $1/(2\Omega\rho \, \sin\phi)$ is constant, but since ρ varies with height so this factor will vary and have a different value for each height considered, and a different geostrophic scale will be required for each height.

Therefore, instead of using charts of pressure at a fixed height above the ground it is preferable to use charts showing the height above mean sea level of a *fixed pressure level*. As these are charts of heights it is appropriate to call them 'contour charts'. There are highs and lows appearing on these charts indicating high heights and low heights of the particular pressure level.

Contour charts are normally prepared for pressure levels of 850 mb, 700 mb, 500 mb, 300 mb, 250 mb, 200 mb and 100 mb, which correspond roughly to mean heights of 5000 ft, 10 000 ft, 18 000 ft, 30 000 ft, 35 000 ft, 38 000 ft, and 54 000 ft. Upper air contour charts are usually drawn with the contour interval at 60 metres for levels up to 300 mb, and 120 metres at 300 mb and above.

It may not be obvious at first sight why contour charts are used. However, the value of the geostrophic wind obtained from them is *independent of air density*, at the level of interest. The contour gradient is analogous to the pressure gradient of a surface isobaric chart. So, just as at the surface the geostrophic wind is inversely proportional to the distance between the isobars for isobars drawn at some fixed millibar interval, so the geostrophic wind obtained from a contour chart is inversely proportional to the distance between the contours for a given fixed interval (the wind blowing along the contour lines).

From a practical point of view, the most important feature of this is that the same geostrophic scale can be used for all pressure levels providing the contours are all drawn with the same height interval (usually 60 m). The direction of the upper winds is obtained just as at the surface by the application of Buys Ballot's law. Wind flow is such that the area of low height is on the left in the northern hemisphere (opposite in the southern hemisphere).

Thickness charts

Charts are also constructed showing the thickness of air between two pressure levels, for example 1000 mb and 500 mb. This provides an additional aid to the difficult problem of assessing the relative intensities of fronts and the future movement and development of weather systems, and an understanding of thickness considerations is of great importance when dealing with the question of why the upper winds and upper air charts are different from those at the surface.

The thickness (or depth) of an air column is proportional to the *mean* temperature of the air between the two pressure levels; thus when we plot a series of thickness contours we are also plotting a series of *mean* isotherms. Also, if we use the same height interval for the thicknesses, we can use the same upper air geostrophic scale to measure the 'thickness geostrophic wind', but since the thickness is also a measure of the mean *thermal* structure of the layer, this wind is more usually referred to as the *thermal wind*. Another variation in definition is that the thermal wind is the vector difference between the geostrophic winds at the bottom and top of the layer, as it considers the temperature distribution in both the horizontal and vertical profiles.

Thus, just as contour charts show the *distribution* of high and low pressure areas, so thickness charts show the *distribution* of warm and cold air masses, and by comparing the two types of charts it is not difficult to see that just as pressure pattern winds flow with low pressure to their left, so thermal winds flow with cold air to their left in the northern hemisphere (opposite in the southern hemisphere). See Fig. 3.1.

Thickness charts are normally drawn for the 1000–500 mb layers. Figure 3.1 shows hypothetical contours of the 1000 and 500 mb pressure surfaces. The *thickness* of the 1000–500 mb layer is proportional to its mean temperature – low thickness values correspond to cold air, high thickness values correspond to warm air. Remember that warm air has 'expanded' giving greater absolute height level for level (for pressure levels) than in cold air.

The theoretical wind vector (V_T) blowing parallel to the thickness lines, with a velocity proportional to the temperature gradient, is termed the *thermal wind*. Remember, the thermal wind is not a 'real wind' but a component. The geostrophic wind velocity at 500 mb (G_{500}) is the vector sum of the 1000 mb geostrophic wind (G_{1000}) and the thermal wind (V_T), as shown in Fig. 3.1.

Since the thermal wind component blows with cold air (low thickness) to the left in the northern hemisphere, when viewed downwind, it is readily apparent that in the troposphere the poleward decrease of temperature should be associated with a large westerly component in the upper winds. Furthermore, since the meridional temperature gradient is steepest in winter, the zonal westerlies are most intense at this time.

The results of the above influences in the northern hemisphere are as follows.

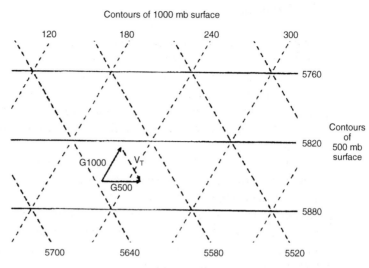

Fig. 3.1 Schematic of superimposed contours of isobaric height and thickness of 1000–500 mb layer (in metres). G1000 is the geostrophic wind velocity at 1000 mb, G500 is the geostrophic wind velocity at 500 mb. V_T is the resultant 'thermal wind' blowing parallel to the thickness lines with low thickness in its left (colder air) (northern hemisphere).

- Most upper geostrophic winds are dominantly westerly between the sub-tropical high-pressure cells (centered aloft about 15° North) and the polar low pressure centre aloft.
- Between the southern flank of the subtropical high pressure cells and the Equator, they are easterly.
- The dominant, westerly circulation reaches maximum speeds of 90–150 kt, which even increase to over 200 kt in winter in some areas. These maximum speeds are concentrated in a narrow band often situated about 30° latitude, between about 30 000 and 50 000 ft. Usually quoted at the 200 mb level. *This is the location of the subtropical jet stream.*

The relation between features of the 500 millibar and surface charts
The 500 mb contour chart can present a useful guide to the location and scale of surface systems. The association between surface and upper air features is discussed with the aid of Figs 3.2 and 3.3 (charts are for March).

(1) A closed cyclonic vortex at 500 mb is usually associated with a surface depression; the systems may be more or less concentric or the surface low may be immediately to the east of the upper vortex. Examples are the vortices over the Norwegian Sea, the Aleutians and Alaska and the vortex south of Greenland. An upper blocking low in low latitudes may have a negligible surface low in association (cold columns of air).

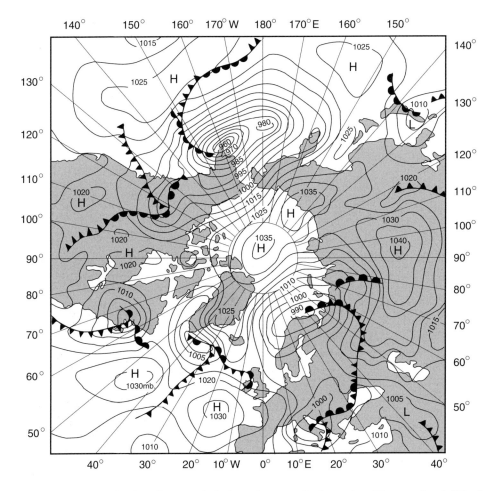

Fig. 3.2 Surface synoptic chart for March, isobars drawn at 5 mb spacing. The associated 500 mb chart is shown in Fig. 3.3.

(2) A trough at 500 mb is associated with a surface trough or depression which will be found directly below or below the eastern flank of the upper feature. Examples are the long-wave troughs over Central Europe, Eastern America and Kamchatka.

(3) Newly developed surface depressions do not necessarily have a significant counterpart in the 500 mb flow. One such feature is the warm front breakaway over Scotland.

(4) A broadly zonal regime at 500 mb is associated with a zonal surface regime. This, however, may be a fronto-genetic zone with incipient wave depressions. Such a regime is shown in the Pacific.

(5) The scale and intensity of the circulation at 500 mb is often directly related to the scale and intensity of the associated surface systems.

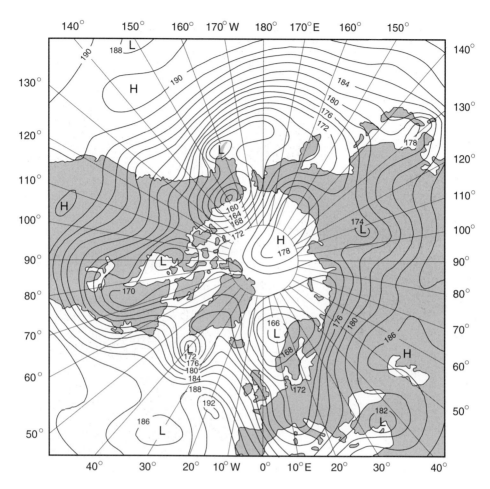

Fig. 3.3 500 mb contour chart for March ($\cong$ 18 000 ft). Contours drawn at 200 ft intervals. H and L are high heights and low heights of the 500 mb pressure surface.

Comparisons of the surface depressions already mentioned (together with their circulations aloft) are of interest in this respect.

(6) A closed anticyclonic vortex at 500 mb is associated with a surface anticyclone, either concentric with or to the east of the upper feature. Examples are the anticyclonic vortices of the eastern Atlantic and Arctic oceans.

(7) An upper ridge at 500 mb is usually associated with a surface anticyclone (or ridge) directly below, or to the east and south-east of the upper feature. Examples are given by the 500-millibar ridges over the Atlantic and America.

(8) Markedly diffluent regions at 500 mb, as at a jet exit region, are associated with a surface depression to the north-west of the jet axis and a surface anticyclone to the south-west. An example is shown by the

diffluent area to the west of the Rockies. Conversely, it may be expected that there are surface anticyclones to the north-west of a confluent area at 500 mb and surface depressions to the south-west.

(9) During the winter season over the continents, it is noted that cyclonic circulations at 500 mb may arise chiefly from cold pools in the 1000–500 mb thickness. In such cases, a marked cyclonic circulation is not necessarily a feature of the surface chart, and indeed at times a cold high may be the preferred surface feature (cold column of air).

3.2 Jet streams

Jet streams are defined by the World Meteorological Organization (WMO) as air currents with quasi-horizontal axes which are thousands of kilometres long, hundreds of kilometres wide, and some 10 000 feet in depth. Wind speeds in the core of a jet stream should exceed 60 kt, with vertical shears of about 10 kt/km and horizontal shears of about 10 kt/100 km. The poleward flowing limb of the Hadley cell produces a significant jet stream over the subtropics at about the 200 mb (39 000 ft) level (see Fig. 3.4). This is known as the subtropical jet (STJ).

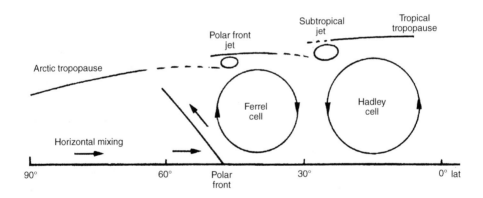

Fig. 3.4 Schematic of meridional circulation showing location of the subtropical jet and polar front jet streams. Northern hemisphere – winter. (After Palmen, 1951).

The position of the subtropical jet stream is clearly marked on upper air charts and it extends almost continuously around the globe, except over Asia during the summer. Since the mean jet stream can be rationalised as a phenomenon arising within the Hadley cell (from the conservation of angular momentum), its position should vary only slightly with the seasons. This can be observed over certain parts of the Earth where the jet stream is only slightly nearer the equator in winter than in summer.

Wind speeds in a jet stream build up to a limiting maximum speed, beyond which the resulting turbulence tends to dissipate excessive kinetic energy in the jet core. In practice, jet stream speeds in excess of 200 kt are unusual and speeds near 300 kt are extremely rare (but have been reported near the western Pacific seaboard). There is a limit to the latitude range over which air flows meridionally from the divergent outflow at higher levels from equatorial latitudes. In the equatorial atmosphere this distance should be between 25° and 35°, since jet streams are found near latitudes 30° North and 30° South.

High level convergence takes place into the jet stream. This is because of internal friction limiting the wind. If the wind speed is limited by turbulence (which dissipates kinetic energy) then this 'converging' air has to go somewhere, and is seen as massive subsidence below the jet. This subsidence in turn leads to the generation of large surface high pressure systems within the subtropics. Divergence at the surface takes place out of the subtropical highs, the equatorial flow forming the return flow of the Hadley cell – the *trade winds*.

In the main, there are two westerly jet streams to understand. The second one is associated with the polar front (see Fig. 3.5) and is called the 'polar front jet stream'. This jet stream results from the steep temperature gradient where polar air and tropical air interact. Note that the subtropical jet stream is related to temperature gradient confined to the upper troposphere. (There is evidence of a belt of very strong westerly winds (the 'polar night jet') in the stratosphere above 50 mb over very high latitudes. This seems to only occur in winter.)

The polar front jet stream is very irregular in its longitudinal location and commonly discontinuous, whereas the subtropical jet stream is much more persistent. The synoptic pattern of polar front jet stream occurrence is further complicated in some sectors by the presence of additional frontal zones, each associated with its own jet stream. This situation is common in winter over North America. Main jet stream cores are associated with the principal troughs of the Rossby long waves.

In Fig. 3.6, typical airflows are shown in relationship to a warm sector depression. It must be remembered that the fronts shown in Fig. 3.6 are located at the surface; however, the warm sector air encompasses a greater area aloft due to the vertical profiles of the fronts. The plan view relationship between the jet stream core and the surface fronts can be seen in Fig. 3.7. The jet stream core is located in the warm sector air aloft. The actual relationship may depart from the idealised case.

Figure 3.7 is related to Fig. 3.6. The airflows show convergence at the cold front where the entrance to the jet stream, 'A', is found. Air is rising, and in Fig. 3.7 the jet stream in this region is seen to be accelerating. The jet stream curves around the surface position of the fronts, and now enters an area of divergence; air is sinking and decelerating. The jet exit is located at 'B'. The height of the jet core is about 1 km higher at the entrance than the exit. Remember too that the jet core is located just below the tropopause. Above a

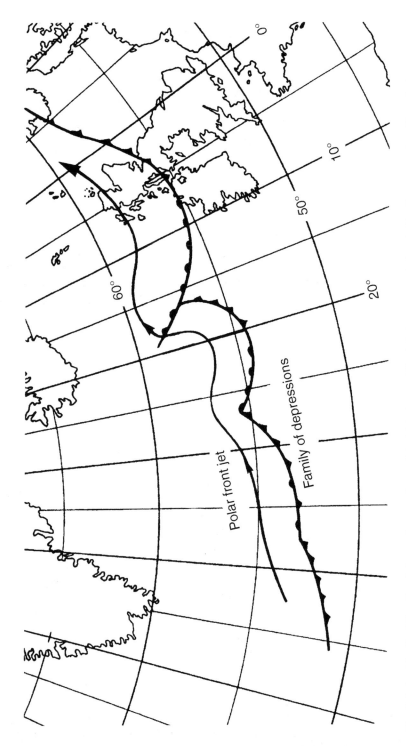

Fig. 3.5 An idealised family of depressions and associated jet stream. The jet stream core is located in the warm air, although drawn on a surface chart it is positioned vertically over the cold air.

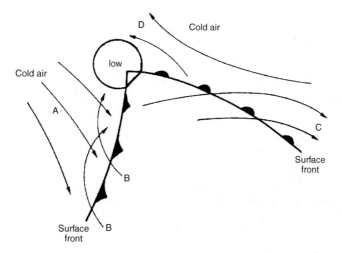

Fig. 3.6 A typical warm sector depression showing the relative air movements. At location A, cold air is moving towards the cold font and subsiding; at location B, low lying warm air is lofted upwards by the cold front; at location C, high level warm air is spilling over the warm front; at location D, cold air is being displaced by warm air.

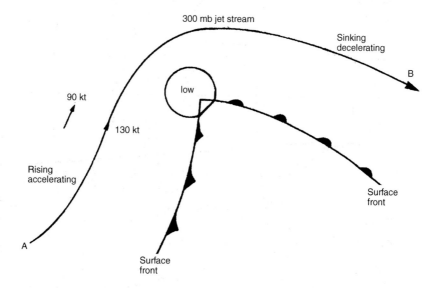

Fig. 3.7 Idealised surface frontal systems and jet stream showing zones of upper tropospheric divergence (rising air) and convergence (sinking air) and also the location of a velocity maximum.

jet stream, there is no temperature gradient as the height is now above the tropopause and the stratospheric temperature becomes isothermal. The wind speed now decreases.

The wind speed along the jet axis varies; its maximum is to the west of the centre of a low pressure system, assuming the low pressure system is active (deepening). Figure 3.7 shows the jet stream maxima in relation to the surface fronts. The 'wave' which forms the warm and cold front moves at about 25 to 45 kt. The warm air in the jet stream is therefore moving at greater speed and overtakes the low-pressure area. It accelerates to a maximum at the jet stream entrance, 'A' in Fig. 3.7 (sometimes called the region of confluence), and decelerates at the jet stream exit, 'B' (sometimes called the region of diffluence). At the area of the jet stream entrance, divergence causes the lower level air to rise on the Equator side (i.e. right-hand side) of the jet stream core, whereas at the jet stream exit, ascent is on the poleward side (i.e. left-hand side).

It has been observed that there is a jet stream with a predominantly easterly flow in equatorial latitudes mainly in those longitudes where the upper easterlies are displaced some 10° or so from the Equator. Note that in all jet streams, the maximum speeds are usually reached just below the level of the tropopause, i.e. the equatorial jet stream at about 100 mb, the subtropical jet stream at about 200 mb and the polar front jet streams at about 250–300 mb.

200 millibar (global) winds

The charts in Figs 3.8 and 3.9 show streamlines of wind direction (dashed lines), and wind speed arrows for the 200 mb level (approximately 38 000 ft). The subtropical jet stream reaches maximum speeds at this level, however it should be noted that polar front jet streams tend to occur at the 300 mb level (approximately 30 000 ft) and the easterly equatorial jet stream at the 100 mb level (approximately 53 000 ft). These charts show the equatorial wind envelope at 38 000 ft and, of particular note, the expanded envelope in July.

From Fig. 3.8, the following points should be noted in respect of the distribution of upper 200 mb winds in January.

(1) The general flow is westerly, however between approximately the longitude of New Zealand, extending through South East Asia, across the Indian Ocean to the longitude of West Africa, there is an easterly envelope, the shaded area.

(2) The subtropical jet stream of the northern hemisphere is found between 25° North and 40° North. It can be seen to extend from North Africa to Japan, passing over the Persian Gulf and India. It reaches its highest mean speed in the region of Japan. (See Chapter 18, 'Weather – Singapore to Japan'). The mean position of the southern hemisphere subtropical jet stream is located at about 40° South.

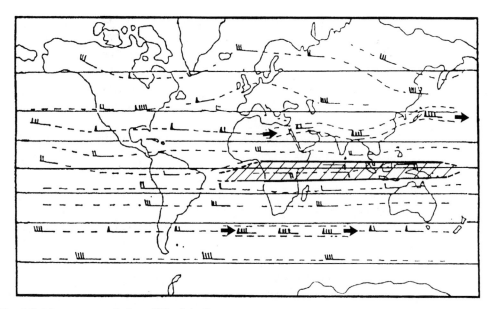

Fig. 3.8 Mean upper winds at 200 mb in January.

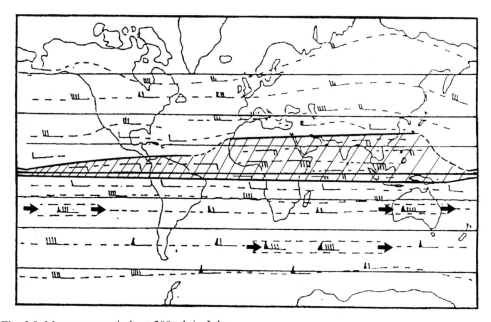

Fig. 3.9 Mean upper winds at 200 mb in July.

(3)   The polar front jet streams are not so apparent on the 200 mb mean wind speed charts, as they are located around the 300 mb level ($\cong$ 30 000 ft). Note: the polar front jet streams change position frequently and, unlike the subtropical jet streams, are not semi-permanent features. The mean 200 mb upper wind over the Atlantic is westerly at approximately 50 kt. Note, however, the strong mean wind speed over the south-east coast of the USA. This is due to the strong temperature gradient created by cold polar air moving across the cold land surface of the USA (with little modification to its temperature) to meet very warm subtropical air from the Azores high in this region.

(4)   No polar front jet streams appear in the southern hemisphere on the January chart because the axis of the jet stream is below the 200 mb level. Furthermore, jet stream activity is reduced as the excursions of cold air from the southern oceans do not extend so far north, and the temperature gradients are therefore reduced.

(5)   Note the 'shaded area' on this chart which shows the easterly wind envelope. The northern boundary is at and just north of the Equator. Air routes from the Gulf to the Bay of Bengal will experience westerly winds.

From Fig. 3.9 the following points should be noted in respect of the distribution of upper 200 mb winds in July.

(1)   In July, with the sun in the northern hemisphere, the northern hemisphere subtropical jet stream has moved north to around 40–45° North. Temperature gradients are now weaker because the land has warmed replacing the very cold air that was present during the winter. The mean wind speed of the subtropical jet stream is diminished.

(2)   The southern hemisphere subtropical jet stream has also moved north with the sun and is now found at 30° South with the highest mean wind speeds being over Australia and the South Pacific.

(3)   The general wind direction away from the Equator is still westerly but now the easterly winds of the Equator extend as far as 20° North over India with mean wind speeds higher than in January, reaching jet stream speeds at the 100 mb ($\cong$ 53 000 ft) level at about 10° North. Note, however, that at normal cruising levels (300–200 mb) the easterly wind at this latitude averages approximately 40 kt.

(4)   The polar front jet stream is less in evidence than in winter but there is still an area of strong mean winds over the eastern USA at around 50° North. In the southern hemisphere the polar front jet stream is in evidence at around 50° South in the South Atlantic and Southern Indian Ocean, but not in the southern Pacific.

(5)   The shaded area showing the easterly upper wind envelope is considerably enlarged as the landmasses to the north are warming up. The air

routes from the Gulf area to Singapore will experience easterly winds. The easterly envelope expands northwards as a result of the landmass warming up during the northern summer. Low pressure now replaces the massive winter high-pressure area, and higher surface temperatures prevail. This leads to the formation of a *thermal high-pressure cell* in the zone from about the 500 mb level and above, which overlays the shallow heat low closer to the surface. This produces an easterly flow on the southern flanks of the upper air anticyclone; the surface low pressure weakens aloft and becomes a high pressure zone in a warm column of air. The 'thermal wind' component is now easterly, and upper winds within the envelope become easterly, and strengthen with height.

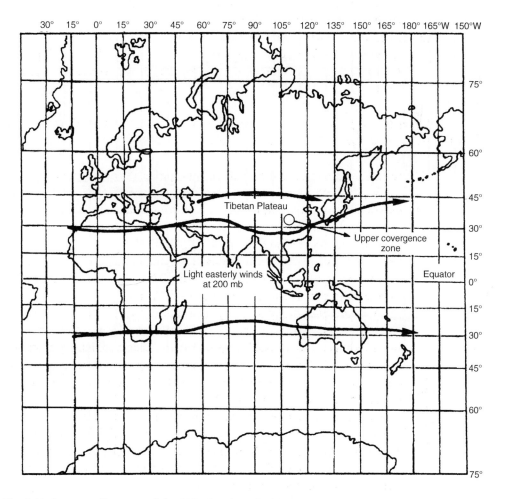

Fig. 3.10 January alignment of the 200 mb subtropical jet streams. The surface winds over India at this time are generally NE (NE monsoon). (After Lockwood, 1965.)

Asian monsoons and jet streams

A note should be made about the Asian monsoon and the associated sub-tropical jet stream. The Asian monsoon is described in Chapter 17. When upper winds are considered, it is found that the Asian monsoon is a complex system. During the northern winter (see Fig. 3.10), the subtropical westerly jet stream lies over Southern Asia, with its core located at $\cong 40\,000$ ft. It divides in the region of the Tibetan plateau, with one branch flowing to the north of the plateau, and the other to the south. The two branches merge to the east of the plateau and form an immense upper convergence zone over China. This is called the *Tibetan lee convergence zone*, and is the mechanism that causes extreme wind speeds.

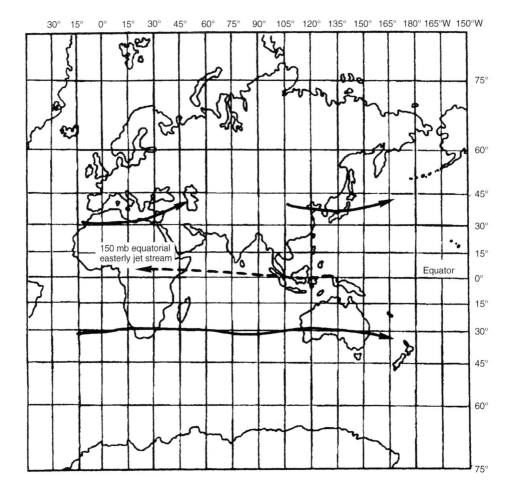

Fig. 3.11 July–August alignment of the 200 mb subtropical jet streams. There is an easterly jet stream at 150 mb (dashed line). The surface winds over India at this time are generally SW (SW monsoon). (After Lockwood, 1965.)

In May and June, the subtropical jet stream over northern India slowly weakens and disintegrates, causing the main westerly flow to move north into central Asia. As noted earlier, the temperature gradients weaken as the landmass to the north becomes warmer. While this is occurring, an easterly jet stream, mainly at $\cong 45\,000$ ft, builds up over the equatorial Indian Ocean and expands westwards into Africa (see Fig. 3.11). The formation of the equatorial easterly jet stream is connected with the formation of an upper-level high-pressure system over Tibet. In October, the reverse process occurs. The equatorial easterly jet stream and the Tibetan high-pressure system disintegrate, while the subtropical westerly jet stream reforms over northern India. Referring to Figs 3.10 and 3.11, it can be seen that the subtropical jet stream in the northern hemisphere is greatly influenced by the land, and in particular mountain ranges, i.e. the Himalayas. In the southern hemisphere, there is a greater degree of uniformity.

References

Lockwood, J.G. (1965) The Indian monsoon – a review. *Weather* **20**, 2.

Palmen, E. (1951) The role of atmospheric disturbances in the general circulation. *Quarterly Journal* of the Royal Meteorological Society **77**, 337.

Chapter 4
Easterly Waves

4.1 Wave disturbances

There are wave disturbances that travel westwards in the equatorial and tropical tropospheric easterlies. Two main areas are affected, the Caribbean and the Philippines. Their wavelength is between about 2000 and 4000 km and they have a lifespan of 1–2 weeks, travelling some 6°–7° of longitude per day. The Caribbean easterly wave has been studied in great detail and will be described. The Pacific easterly wave is of a similar structure.

Dealing with the Caribbean easterly wave, observations indicate that waves form in the easterly flow over the Caribbean, and that they move west at a speed of 10–15 kt. The structure of such an easterly wave is shown in Fig. 4.1.

Waves are typically about 15° latitude across, and the easterly winds flow through them from east to west. The wave structure in the easterlies is found to be weakest at sea level, to increase in intensity up to about 13 000 ft and then above this level to again become weaker.

About 300 km ahead of the wave trough, the trade-wind inversion is at its lowest level and exceptionally fine weather prevails; this is an indication of intense subsidence. A rapid rise in the altitude of the trade-wind inversion and therefore an increase in the depth of the surface moist layer takes place near the trough line. (This can be clearly seen in Fig. 4.1b.) The moist layer attains a maximum depth of well above 20 000 ft in the area where lofting takes place behind the trough. In this region are found large squall lines, intense rain, and rows of cumulonimbus clouds. Interestingly, the cumulonimbus clouds are not spaced as found in air mass instability, often being organised into lines or rows and there are wide zones of subsidence between individual cloud lines.

Easterly waves, when established, move towards the west at a speed slightly less than the prevailing trade wind. At this stage, evidence of the advancing trough line sees the wind backing and cloud increasing behind the trough in amount and particularly height (main cloud band). The amplitude of the wave pattern is greater at about 13 000 ft, which bears out the earlier statement that the system intensifies with height, then decreases further aloft.

In a deep easterly wave system, pressure can drop in the centre to as low as 1000 mb and form a closed vortex. At this stage, the cloud patterns change, as

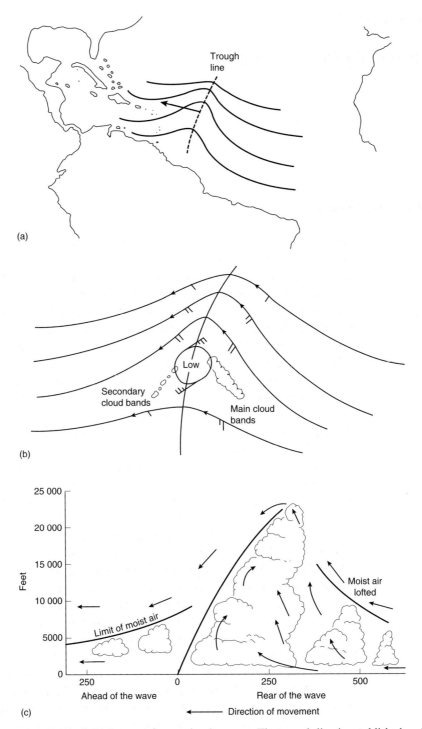

Fig. 4.1 (a) The initial stage of wave development. The trough line is established and cloud is building up at the rear. System tracks westwards at the southern margin of the high pressure belt. (b) The easterly wave fully developed, with secondary cloud bands ahead of the trough line. In the centre, an enclosed vortex. (c) A vertical section through an easterly wave. The trade-wind inversion is lofted up particularly behind the trough line (Figs 4.1a–c from Malkus, 1962).

cloud is now developing ahead of the centre of the low (secondary cloud band). Furthermore, additional cloud bands can form in the general area behind the trough line.

In Fig. 4.1b a vertical section indicates the trade-wind inversion and that the height of the moist air (trade-wind inversion) is quite low ahead of the system which is subjected to descending drying air. However, it is gaining height ahead of the trough line. There is marked instability immediately behind, and the moist air that was lofted to $\cong 15\,000$ ft or so can now be seen to descend to resume a normal level after this disturbance has passed through.

These waves are quite unlike a mid-latitude depression. Note that the cloud is essentially behind the trough line, and slopes eastwards with height. The pattern is associated with horizontal and vertical motion of the easterlies. Behind the trough, low level air undergoes convergence, while ahead of it there is divergence.

The passage of such a wave produces the following weather sequence:

(1)   In the ridge ahead of the trough, fine weather, scattered cumulus cloud, some haze.
(2)   Close to the trough line, well-developed cumulus, occasional showers, improving visibility.
(3)   Behind the trough line, veer of wind direction, heavy cumulus and cumulonimbus, moderate or heavy thundery showers and a decrease of temperature.

The surges of the trade wind in the northern China Sea are similar to easterly waves, but there is no record of persistent travelling waves in the China Sea. Though the origins of waves are often obscure, it appears that many over the Caribbean have their sources over Africa. (This will be explored later.)

Reference

Malcus, J.C. (1962) Large-scale interactions. *The Sea*, Vol. 1. John Wiley, New York.

Chapter 5

The Inter-tropical Convergence Zone (ITCZ)

5.1 Introduction

The ITCZ* (or Equatorial trough) lies within the tropics and marks the rather broad zone of separation between air masses conveyed by the trade winds from areas on opposite sides of the Equator. This zone is described as a source region, but air originating in this region has a surface excursion limited by the north/south movement of the ITCZ, and the boundaries of the permanent high-pressure belts. There are, however, modifications to this arrangement by monsoon winds, particularly those of Asia.

This convergence zone in former times was referred to as the 'inter-tropical front' or ITF, but this term was hardly appropriate, particularly after more modern methods of observation and analysis have since been implemented. Temperature and density differences across it are generally small and it has little in common with the frontal systems of higher latitudes. Furthermore, frontal mechanisms similar to those in temperate latitudes are not possible within about 5° of the Equator (due to the absence of the Coriolis force).

Climatologists did not favour the ITF description because clouds could be very irregular, or continuous (see Fig. 5.4) or even absent for hundreds of kilometres. The movement of clouds laterally could take place in a continuous manner, and at other times could move north or south without affecting the intermediate latitudes. This variation of movement and characteristics bear no relation to temperate latitude fronts. Therefore, the concept of a zone has been accepted as more appropriate.

Its position normally varies little from day to day but there are changes in its activity and hence in cloudiness along its length. The ITCZ undergoes a regular seasonal shift in its position during the year (see Figs 5.2 and 5.3), p. 64 and this will be described in more detail.

Carried along the flow of the trade winds, the trade-wind inversion

---

* *Note:* In some modern texts, the ITCZ is described as the 'Inter-Tropical Confluence' (ITC). In this text, ITCZ will be retained.

61

enters the equatorial trough; however, the height of the inversion increases as it nears the Equator and invariably disappears within the trough as vertical displacement of air forming the towering clouds breaks down the inversion.

Vertical development of clouds is restricted in the trade wind, but close to the Equator towering Cu and Cb clouds give heavy showers. This zone then can be described as relatively wet and cloudy.

It must be pointed out however that the zone is not an area of continuous cloud and rain. As large areas are dominated by the trade winds, the trade-wind inversion can produce areas approaching desert-like conditions. Clouds and rain seem to favour certain localities. The big rainstorms seem to form in rather an irregular manner, and once formed either remain stationary or tend to drift west in the equatorial flow.

The ITCZ can exhibit single, double or multiple structures. Bifurcation of the ITCZ is a common feature (see Fig. 5.1, where bifurcation is illustrated), and the boundaries are described as 'shear lines', both north and south.

Trade winds are generally confluent into the ITCZ regions but occasionally winds with a westerly component separate the shear lines. Cyclonic circulations as shown in Fig. 5.1 may form on the shear lines. Assuming that the ITCZ is near the Equator and the circulations around the depressions are drawn in the sense appropriate to the particular hemisphere, then the ITCZ shows air moving into the zone (convergence). This low-level air moves into the zone quicker than it can move away (horizontally) and as a consequence some air must ascend.

Where there is divergence downstream, descending air makes good the deficiency of horizontal inflow. The convergent parts of the ITCZ and the cyclonic circulations set up on the shear lines are of course features of the equatorial trough found on the general circulation model. The areas of cyclonic circulation are the breeding areas of tropical cyclones, should all other factors be present (see Chapter 6). The shear lines and the ITCZ when actively convergent with strong upward motion are zones or belts of cloud, frequently including cumulonimbus, heavy rain, thunderstorms and squalls. However, in the areas where subsidence is involved they are often marked by little more than a few oktas of low cloud.

Across the ITCZ, the width of the belt of disturbed weather varies according to the scale of the convergence. The width may range from about 50 to 500 km. It has been generally observed that when the ITCZ is wide, it is quite active, and when it is narrow, it is less so. Moreover, the leading edge of the ITCZ when moving can also show considerable activity. Outside the clouds, visibility is good, except in heavy rain where it may be reduced to a few metres. The cloud base is usually 1000 ft or more above the sea but it may descend practically to the surface in very heavy rain.

The surface wind in the vicinity of the line of disturbance is squally, often reaching Beaufort force 5 ($\cong$ 20 kt), while squalls of force 8 or 9 ($\cong$ 35–45 kt

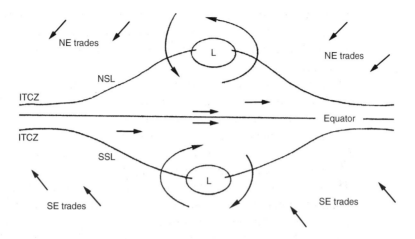

Fig. 5.1 Local spreading (bifurcation) of the ITCZ forming northern and southern shear lines. Notice the low pressure areas forming on the shear lines and rotation appropriate for the respective hemisphere. The surface winds are predominantly westerly within the bifurcation. The distance across the bifurcation at its widest point can be > 500 km.

with gusts of between 43 and 60 kt) are occasionally reported. On occasions of vigorous development, some large rain areas occur and thick masses of nimbostratus and altostratus with embedded cumulonimbus are found, with violent turbulence within the clouds. Usually, however, conditions are somewhat patchy and there are areas of lighter rain and broken cloud.

The height of the cloud tops in the vicinity of the lines of disturbance varies considerably. On many occasions the belt of cloud may be flown over without difficulty at $\cong 20\,000$ ft but on occasions of active development, cloud should be anticipated well above that level, even up to or just above the tropopause, which is in the neighbourhood of $\cong 55\,000$ ft in tropical regions. The spreading out of the cloud tops into extensive sheets of cirrus occasionally conceals the cumulonimbus turrets with their severe turbulence. At other times the cirrus sheets may persist long after the convective cloud has dispersed. Ice accretion in active areas is liable to be serious in convective cloud at temperatures from $0°C$ ($\cong 16\,000$ ft) to $-40°C$ ($\cong 35\,000$ ft) or even lower temperatures. This height range is somewhat higher ($\cong 2000$ ft higher) over southern Asia during the northern summer.

Observation of satellite imagery indicates that inter-tropical convergence zones can only rarely be identified as long, unbroken bands of heavy cloudiness. Rather, they are usually made up of a number of what climatologists describe as 'cloud clusters'. These are usually separated by large expanses of relatively clear skies. There is strong evidence that cloud clusters are, in turn, manifestations of synoptic-scale westward propagating wave disturbances, which are marked by heavy precipitation.

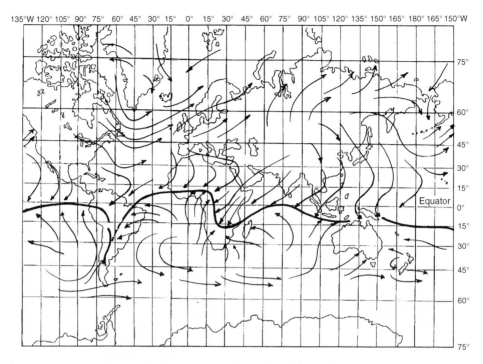

Fig. 5.2 Prevailing surface winds in January and location of the ITCZ.

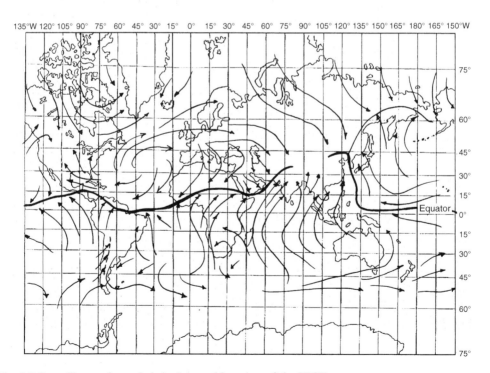

Fig. 5.3 Prevailing surface winds in July and location of the ITCZ.

Cloud clusters represent concentrations of cyclonic circulation at low levels and anticyclonic circulation at the 200 mb level, with strong ascent at middle levels. The clear areas between clusters are marked by weak subsidence together with vertical distributions of circulation and divergence opposite in sign to those within the clusters.

In Fig. 5.2 the prevailing surface winds for January are shown, also the confluence denoting the ITCZ. Notice that the general line of the ITCZ during January has an enormous excursion southwards over the major landmasses. There are two significant splits illustrated, one over South America and the other over north of New Guinea.

In Fig. 5.3, the prevailing surface winds for July are shown. This time the ITCZ has moved north, with the exception of the western Pacific, but the main

Fig. 5.4 Satellite image. Visible light spectrum, format C4D, Atlantic and South America. The ITCZ is shown as an almost unbroken line at about 10°N. It is positioned at its further northern excursion, summer (July). (Reproduced with permission of ESA/Eumetsat.)

excursions can be seen over major landmasses, particularly the east coast of China, and into the Sea of Japan. There is a break in the general line of the ITCZ over the Tibetan plateau, as its continuation is ill-defined, with little or no temperature gradient over the land. See Fig. 5.4 which shows a satellite image of a well-marked ITCZ across the Atlantic in July.

Chapter 6

Tropical Storms

A closed circulation system producing storms outside the latitude bands of 5° North to 5° South can occur at any time within the tropics. The storms can assume slowly circulating masses of air with scattered cumulonimbus clouds to very violent disturbances. Differences between the different tropical low-pressure systems are not easily defined. The World Meteorological Organization has classified low-pressure systems as follows.

(1) A tropical depression is a system with low pressure enclosed within a few isobars, and it either lacks a marked circulation or has winds below 33 kt.
(2) A tropical storm is a system with several closed isobars and a wind circulation from 33 to 63 kt.
(3) A tropical cyclone is a storm of tropical origin with a small diameter (some hundreds of kilometres), minimum surface pressure less than 900 mb, very violent winds, and torrential rain accompanied by thunderstorms. It usually contains a central region, known as the 'eye' of the storm, with a diameter of the order of some tens of kilometres, where there are light winds and more or less lightly clouded sky.

Tropical cyclones are given a variety of regional names. In the south-west Pacific and Bay of Bengal the name *tropical cyclone* is in use; they are known as *typhoons* in the China Sea, *hurricanes* in the West Indies and *cyclones* in the South Indian Ocean. Plentiful moisture is a requirement of tropical cyclones, and for this reason their incidence is limited to regions where the highest sea-surface temperatures are found ($> + 27°C$), that is to the western regions of the tropical oceans in the late summer.

In the central equatorial oceans, the trade wind systems of the two hemispheres converge in the Equatorial trough and wave disturbances may be generated if the trough is sufficiently removed from the Equator to provide a small Coriolis force to initiate a vortex. These disturbances quite often become unstable forming cyclonic rotation (rotation direction dependent on hemisphere) as they travel westwards attaining greater strength. The winds do not necessarily all attain hurricane strength, and not all low-pressure systems become tropical cyclones; many remain as weak closed circulations with moderate rain and light winds.

Convergence in the Equatorial trough can create a vortex setting up the initial stage required in forming a tropical storm. In Fig. 6.1 the ITCZ is drawn as a line showing the leading edge of the zone advancing towards the north, known as the northern shear line (NSL). It can be seen that on passing north of the Equator, the south-east trade winds swing round to the west or south-west as the Coriolis parameter becomes more effective, although within the boundaries of the Equatorial trough, the winds are predominately westerly.

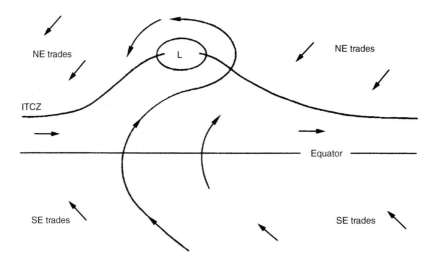

Fig. 6.1 When the ITCZ migrates away from the Equator, any waves which form can produce a vortex as the Coriolis parameter is increasing. The direction of rotation is dependent on hemisphere.

If a wave forms on the ITCZ it can cause a circulation or vortex, and the surface streamlines may continue around this wave where pressure is beginning to drop. Convergence is taking place, and air starts to ascend. If all parameters are in place (as now described) a tropical cyclone is a possible consequence. The mechanism in the Equatorial trough is described in greater detail in Chapter 5.

Figure 6.2(a) and (b) shows the cross-section and plan of a fully developed tropical cyclone. As the structure shows, ascent within tropical cyclones takes place in cumulonimbus clouds arranged in spirals converging on the central eye. The spirals, which are sometimes hundreds of kilometres long, are at most a few kilometres wide, and the distance between them is about 50 to 80 km near the edge, decreasing towards the central eye. This means that often only a small fraction of the tropical cyclone (not more than 10%) contains the ascent which gives rise to the main contribution to condensation and rainfall.

Close to the centre, where the clouds form a ring around the eye, the

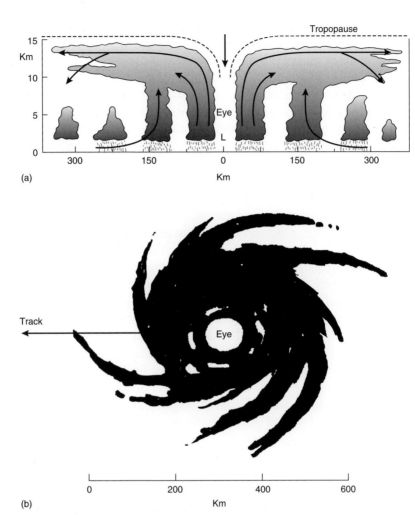

Fig. 6.2 (a) Schematic vertical profile of a tropical cyclone/hurricane of moderate strength. It consists of a central eye surrounded by Cb towers. Air descends within the eye, whereas considerable vertical updrafts are present in the Cbs forming the eye wall. (b) Plan view of a tropical cyclone/hurricane. Cbs are clustered around the eye and radiate outwards in spirals.

strongest winds and heaviest rainfalls combine to produce the storm's full force, whereas inside the eye the winds decrease quickly and the heavy rains cease. Research has confirmed that the eye is actually a region of subsiding warm air at the centre of the storm. This warm core seems to be a characteristic and essential to the formation of a tropical cyclone, and its appearance is one of the first signs that a storm is going to turn into a cyclone.

The typical cyclone system has a diameter of about 600 km, which is smaller (about half the size) than that of a mid-latitude depression, although typhoons in the China Sea are often much larger. The central pressure at sea level is commonly 950 mb and exceptionally falls below 920 mb. Lowest ever recorded in the western hemisphere was 888 mb, Hurricane Gilbert 1988! Hurricane winds are defined arbitrarily on the Beaufort wind scale as 64 kt or more, and in many storms they exceed 100 kt and some reach 200 kt. *The storm as a system moves at a speed of about 15 kt.*

The extensive vertical development of cumulonimbus cloud with tops at over 14 000 m ($\cong$ 45 000 ft) indicates the immense convective activity found in such systems. Radar and satellite studies show that the convective cells are normally organised in bands which spiral inwards towards the centre (Fig. 6.2b).

A number of conditions are observed to be necessary for cyclone formation. One requirement is an extensive ocean area with a surface temperature greater than 27°C. Cyclones rarely form near the Equator, where the Coriolis parameter is close to zero, or in zones of strong vertical wind shear (i.e. beneath a jet stream), as both factors inhibit the development of an organised vortex.

There is also a definite connection between the seasonal position of the Equatorial trough and zones of cyclone formation, which is borne out by the fact that no cyclones occur in the South Atlantic (where the trough never lies south of 5°S) or in the south-east Pacific, east of about 140°W (where the trough remains north of the Equator, and its a 'cold water' coast) (see Chapter 2, section 2.4). On the other hand, satellite imagery from over the north-east of Panama shows an unexpected number of cyclonic vortices in summer, many of which move westwards near the trough line about 10°–15°N.

About 60% of tropical cyclones seem to originate 5°–10° latitude poleward of the Equatorial trough in the doldrums sector, where the trough is at least 5° or so from the Equator. The development regions of cyclones lie mainly over the western sections of the Atlantic, Pacific and Indian Oceans where the subtropical high-pressure cells do not cause subsidence, and stability and the upper flow is divergent, and the sea is warmer.

Theories about cyclone development in former times believed that convection cells generated massive release of latent heat to provide energy for the storm. Although convection cells were regarded as an integral part of the hurricane system, their scale was thought to be too small for them to account for the growth of a storm hundreds of kilometres in diameter. Research, however, reveals that energy is apparently transferred from the cumulus-scale to the large-scale circulation of the storm through the organisation of the clouds into spiral bands (see Fig. 6.2b). In other words, the organisation of individual cumulonimbus cells is acting together as one very big system.

The enormous quantity of heat liberated in the tropical cyclone makes the temperature aloft higher than that of the surrounding atmosphere. The thermal wind is thus anticyclonic such that the cyclonic flow decreases with height in the lower troposphere with an anticyclonic flow at still higher levels. The divergence associated with this anticyclonic outflow is the cause of the cirrus extending beyond the main cloud mass.

It follows that as the thermal wind becomes anticyclonic aloft, and there is cyclonic circulation at lower levels, the wind direction will reverse. This is particularly so on the advancing side of the storm. For a northern hemisphere storm as shown in Fig. 6.2b, the highest surface winds would be in the northwest quadrant, where the storm speed (of about 15 kt) is added to the circulation speed. However, aloft at $\cong$ the 200 mb level, the winds become southerly in this quadrant radiating from the eye, and then follow a spiral trajectory as shown by the cloud patterns in Fig. 6.2b. In other words, the spirals at these levels are *outward* flowing, entraining the high level cirrus cloud and following the thermal vector, whereas the central lower part of the storm obeys cyclonic rotation. These counter flows are readily seen on satellite time-lapse images.

There is now evidence to show that hurricanes form from pre-existing disturbances (see caption with Fig. 6.3 – Hurricane Andrew), but while many of these disturbances develop as closed low-pressure cells few attain full hur-

Fig. 6.3 A satellite image of Hurricane Andrew – NOAA HRPT, August 1992. Hurricane Andrew was first tracked as a small depression off the west coast of Africa and took about 12 days to reach and cross southern Florida. This image shows the storm just about to leave the west coast. It entered the Gulf of Mexico on 23/24 August and continued across the Gulf to hit Louisiana on 26 August. Southern Florida was declared a disaster area. This was the worst storm for 60 years. (Acknowledgements to NOAA, photo © of Timestep supplied by Dave Cawley and reproduced with permission.)

ricane intensity. The reason for this appears to be the presence of an anti-cyclone in the upper troposphere. This is essential for high-level outflow, which in turn allows the development of very low pressure and high wind speeds near the surface.

It can be seen in Fig. 6.2(a) (vertical profile) that there is a counterflow near the tropopause. The anvil tops of the cumulonimbus clouds indicate out-flowing (divergent air movement) which then follows the radial spirals. Above this level, there is an in-flowing (convergent air movement – high pressure) which feeds into the 'eye' producing the marked warm descending air column. The dynamics of this mechanism is not entirely understood, and is the subject of ongoing investigation by the atmospheric physicists.

A distinctive feature of the hurricane is the warm vortex (eye), since other tropical depressions and incipient storms have a cold core area of shower activity. The warm core is enhanced by the action of up to 200 cumulonimbus towers releasing latent heat of condensation. At any one time about 15% of the area of cloud bands is giving rain. Observations show that although these 'hot towers' form less than 1% of the storm area within a radius of about 400 km, their effect is sufficient to change the local environment.

The warm core is vital to hurricane growth because it intensifies the upper anticyclone, upper air convergence, and this leads to a 'feedback' loop by stimulating the low-level input of warm moist air which further intensifies convective activity and latent heat release, with consequential upper-level high pressure.

In the eye, or innermost region of the storm, adiabatic warming of descending air accentuates the high temperatures, although since high temperatures are also observed in the eye-wall cloud masses, subsiding air can only be one contributory factor. The eye has a diameter ranging from a few kilometres to over 100 km, but 30–50 km is the usual diameter. Observations of hurricanes in the Caribbean area have seen the intensity of storms increase with a decrease in the eye diameter, down to 10 km in some examples, and these have been particularly intense. Within this region the air is virtually calm and the cloud cover may be non-existent, or broken.

If the rotating air conserved absolute angular momentum, wind speeds would become infinite at the centre and clearly this is not the case. The strong winds surrounding the eye are more or less in cyclostrophic balance, with the small radial distance providing a large centripetal acceleration.

There is an enormous pressure gradient across the storm system, and air is forced to rise when the pressure gradient can no longer force it further inward, the rising air forming the ring of cumulonimbus towers. It is possible that the cumulonimbus anvils play a vital role in the complex link between the hor-izontal and vertical circulations around the eye. The redistribution of angular momentum is thought to be the mechanism behind the maintaining of storm intensity by concentrating the rotation near the centre. Storm intensity is maintained by the continuing supply of heat and moisture combined with low frictional drag at the sea surface, the release of latent heat through condensation, and the removal of the air aloft, which are essential conditions.

As soon as one of these conditions changes the storm decays. This can occur quite rapidly if the track (determined by the general upper tropospheric flow) takes the vortex over a cool sea surface or over land. On passing over land, the increased friction hastens the process of filling, but the cutting-off of the moisture supply removes one of the major sources of heat. Rapid decay also occurs when cold air is drawn into the circulation. Also if the upper-level divergence pattern moves away from the storm, less air is removed from aloft and the mechanism begins to break down.

Once formed, storms move initially in a westward direction (see Fig. 6.5) along the equator side of the subtropical high-pressure margins. In general, they 'recurve' towards the nearer pole as they reach the western edges of the high-pressure areas. That is to say, the storm now recurves towards the east, and will continue to do so if still over the sea. However, in the south-west Pacific region, only about 50% of storms actually re-curve in this manner. (See Chapter 20, 'Weather in the South-west Pacific Region').

At higher latitudes, the storm moves into the westerlies. A further aspect to consider is that as the storm moves further away from the equatorial region, the Coriolis force is increasing. It has been observed that the circulation at higher latitudes becomes less intense, and the centripetal acceleration also decreases, the balance of forces changing as the Coriolis force increases, balancing (or attempting to balance) the pressure gradient.

The system now changes its character becoming a cold core system if still over the sea, which is colder at higher latitudes, further contributing to weakening of the system. The storm could engage polar air to continue as an extratropical depression generating a frontal system and spreading in size. Although it has been mentioned that the system is weakening, this is relative. The original storm was very powerful, but despite 'weakening' it can exhibit considerable bad weather characteristics at higher latitudes.

There are occasions when a depression reaching the British Isles has been traced back to a tropical origin. These depressions can produce high winds when they reach Europe; what is more likely is that the warm tropical air which is retained aloft produces a greater than average (heavy) rainfall.

The main hurricane (and typhoon) activity in the northern hemisphere is in late summer and autumn during the time of the Equatorial trough's northward displacement. The 'hurricane season' occurs in the western Atlantic mainly between *May* and *November* with a marked peak in September (when the Equatorial trough starts to move south again), and in the western Pacific between *July* and *October*. A small number of storms may affect both areas as early as May and as late as December. The late summer–autumn maximum is also found in the other areas, although there is a secondary, early summer maximum in the Bay of Bengal. These secondary maxima seem to relate to the Equatorial trough when it is in

Fig. 6.4 A tropical cyclone, its centre positioned over the Gulf of Kutch – north-west India. This is a rare event in this ocean and even rarer on this side of the Arabian Sea. Image as seen by Meteosat 0900 UTC 9 June 1998. (Photo supplied by Peter Wakelin RIG. Acknowledgement to EUMETSAT.)

motion. In Fig. 6.4 can be seen a satellite image of a tropical cyclone in the Arabian Sea, although rare in this particular position. More often they are seen further south and west.

The cyclone season in the *south-west Indian Ocean* officially spans from 1 November to 15 May. Each year, on average, 10 depressions reach at least moderate intensity (wind force 8 on the Beaufort scale: 34–40 kt) at which stage they are given a name. Of these, three intensify into tropical cyclones (wind force 12 on the Beaufort scale: > 64 kt). January and February are the most active months and account for half of the storms.

The 1997–98 season started exceptionally late and was very quiet with only seven named depressions. Most of these were either short lived or just reached moderate intensity. Only *Anacelle* (the first depression) intensified into a tropical cyclone.

Since records started about 150 years ago there have been only a few years when activity started as late as during the 1997–98 season. However, it was not until the advent of meteorological satellites in the 1960s that all tropical depressions could be detected and their intensity and tracks determined with fair accuracy. The records for the last 30 years are therefore more reliable. Over that period, the 1997–98 season stands out as the only one with no named depression until February.

Should El Niño be blamed for this anomaly? An examination of the records indicates that during previous warm El Niño episodes, cyclone activity has at times been above and at times below average. For instance, during the 1972–73 summer there were 12 depressions, of which 6 intensified into tropical cyclones, whereas during 1986–87 only 5 depressions were named and none reached the tropical cyclone stage.

It is worth noting that during the 1982–83 El Niño episode, which until the 1997–98 event had been the strongest on record, cyclone activity was similarly low with only six tropical depressions, but in that case the season started in October and ended all of a sudden in mid-January, whereas the 1997–98 season did not start until the second week of February.

Therefore it seems difficult to establish a direct relationship between El Niño and tropical cyclone activity in the south-west Indian Ocean basin. It should however be noted that very few of the major cyclones which have affected Mauritius over the last century occurred during El Niño years. It is also satisfying to note that none of the Mascareignes Islands or Madagascar was affected by a major cyclone during the 1997–98 season, a fact which is unfortunately rather rare (Pougnet, 1998). See Chapter 28 for further details about the El Niño.

Annual frequencies of tropical cyclones are shown in Table 6.1. The figures are only approximate since in some cases it may be uncertain as to whether or

Table 6.1 Annual frequencies of tropical cyclones (maximum sustained wind speeds exceeding 50 kt).

Location	Annual frequency
Western north Atlantic	10
Western north Pacific	27
Eastern north Pacific	14
Northern Indian Ocean	5
Northern hemisphere total	56
Western south Pacific	8
South-west Indian Ocean	7
South-east Indian Ocean	14
Southern hemisphere total	29
Global total	85

not the winds actually exceeded hurricane force and storms in the more remote parts of the South Pacific and Indian oceans frequently escaped detection prior to the use of weather satellites.

Recurvature of cyclone tracks in the southern hemisphere is in the opposite sense, storms moving first towards the west or south-west and later towards the south-east, but they never cross the Equator. Notice also from Fig. 6.5 that no storms appear in the South Atlantic south of the Equator, or in the eastern Pacific south of the Equator.

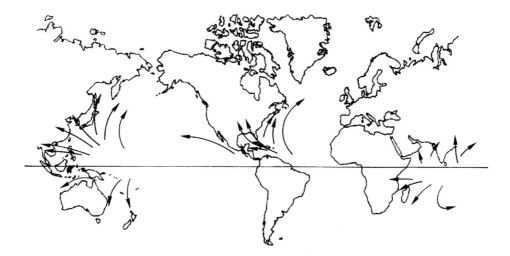

Fig. 6.5 The main tracks of tropical cyclondes.

6.1 Classification of tropical cyclones

Tropical cyclones are classified in accordance with the World Meteorological Organization's recommendation by their maximum wind speeds near the centre. In Hong Kong, the classification is defined in terms of wind speeds averaged over a period of 10 minutes (see Table 6.2).

Table 6.2 Classification of tropical cyclones.

Classification	Maximum 10-minute mean wind near the centre
Tropical depression	up to 62 km/h
Tropical storm	63 to 87 km/h
Severe tropical storm	88 to 117 km/h
Typhoon	118 km/h or more

Saffir–Simpson scale

This is a   hurricane intensity scale used by the United States National Weather Service for assessing the damage that is likely from wind and storm surge from a hurricane. The scale ranges from 1 (minimal) to 5 (catastrophic). See Table 6.3.

The US NWS also designate names to hurricanes when they reach class 1. The first letter of the names chosen are sequential throughout the hurricane season.

Table 6.3 Saffir–Simpson scale.

Type	Category	Pressure (mb)	Winds		Surge (ft)
			Kt	mph	
Depression	TD	–	< 34	< 39	–
Tropical storm	TS	–	34–63	39–73	–
Hurricane	1	> 980	64–82	74–95	4–5
Hurricane	2	965–980	83–95	96–110	6–8
Hurricane	3	945–965	96–112	111–130	9–12
Hurricane	4	920–945	113–134	131–155	13–18
Hurricane	5	< 920	> 134	> 155	> 18

The Dvorak technique

(Personal communication from Jacques Pougnet.)

This is the generally accepted method of estimating tropical cyclone intensity using satellite imagery. It is based on cloud features and their distribution around the storm. Several tropical storm patterns have been established and parameters, such as the size of spiral bands, the diameter and embedded distance of the eye, are considered. The analysis makes it possible to classify the storm on an intensity scale, by giving it a T number.

The intensity scale is non-linear and goes up by steps of half numbers from 1 to 8. According to the model, a typical storm intensifies by one T number per day. The T number is adjusted to take into consideration situations such as re-intensification or weakening and a Current Intensity (C.I.) number is determined. The C.I. number gives the best estimate of the current maximum wind speed and central pressure of the storm.

This relationship was empirically derived from actual observations in the Atlantic basin. The relationship between Dvorak's C.I. numbers and highest sustained winds and lowest sea-level pressure has been modified to reflect conditions in other ocean basins. Intensities (T numbers) of 4.5 or less relate to tropical storms or depressions and those of 5.0 or more to tropical cyclones. C.I. 7 cyclones would generate maximum gusts of the order of 170 knots

(315 km/h) and have a central pressure of about 900 mb. Fortunately, tropical cyclones of such intensity are relatively rare.

The complete classification of tropical storms in the south-west Indian Ocean area, together with the relationship between the C.I. number and wind speed and central pressure is published by the WMO (1998).

References

Dvorak, V.F. (1984) Tropical cyclone intensity analysis using satellite data. Report issue 11: 1984. NOAA, NESDIS technical.

Pougnet, J. (1998) The 1997–1998 cyclone season in the south west Indian Ocean. *Journal of Remote Imaging Group* **53**, 58–60.

WMO (World Meteorological Organization) (1998) Tropical cyclone operational plan for the south west Indian Ocean. WMO (WMO/TD No. 577, Report No. TCP-12).

Chapter 7
Upper Air Temperature and Tropospheric Heights

The temperature lapse rate throughout the troposphere averages 2°C per 1000 ft, but in the lower stratosphere the temperature increases or changes very little with height (it is isothermal in the standard atmosphere). What are of particular importance in air navigation are the heights of the 0°C and −40°C isotherms. The importance of these temperature limits is that they are the upper and lower temperature excursion within which most types of airframe icing is experienced. Moreover, the height of the tropopause is also required.

7.1 Height of the 0°C isotherm

This height is important as it is directly related to airframe icing. The distribution of the mean height over the world is shown in Fig. 7.1 for January and Fig. 7.2 for July. Notice that near the equator it is about 16 000 ft, while in July it increases height to about 18 000 ft in some areas of southern Asia.

As the North Pole is approached, the height decreases, in the first instance as latitude is increased, but it soon exhibits wide variation at given latitudes because of the variation of land/sea surface. This is particularly noticeable over Siberia and North America in January. In the southern hemisphere, this wide variation is not so evident as the landmass is considerably less, and therefore does not present large variations of surface temperature.

7.2 Height of the −40°C isotherm

Airframe icing at these temperatures is rare. In January, the mean height decreases from about 35 000 ft over the Equator to about 16 000 ft over the North Pole. In July, the heights range from about 38 000 ft (over north-west India) to approaching 25 000 ft over the Arctic.

The mean temperature charts can indicate certain significant features, such as the temperature gradients shown by the crowding together of the iso-

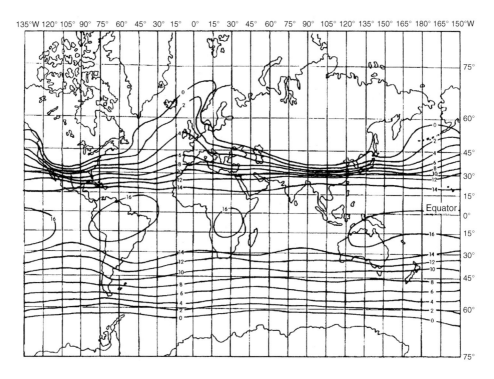

Fig. 7.1 Height (in thousands of feet) of the 0°C isotherm in January. The effect of the warm Gulf Stream drift can be seen clearly to the north of the British Isles.

therms (steep gradient) which occur in the winter on the eastern side of the continents particularly in the northern hemisphere. From Fig. 7.1 the steep temperature gradients can been seen over North America and north of the Himalayas/Tibetan plateau. This 'bunching up' of the isotherms is termed an *isotherm ribbon*, where the temperature gradient is many times greater than normally found in the atmosphere. The effect of these temperature gradients is to see a greater frequency of depressions and intensities in winter than in summer.

There is a marked difference in the alignment of the isotherms across North America and across Europe and Asia in summer. From the Pacific Ocean to the North American continent there is a rapid rise in mean temperature, and the warmest part of the continent is located in the west. From the Atlantic Ocean across Europe to Asia the change is very gradual, and the highest summer temperatures are in the east.

The reasons behind this temperature distribution can be seen with the prevailing winds and the topographical arrangement of mountain barriers. There is a mountain barrier (Rocky Mountains) in the path of the cool maritime winds from the Pacific into mainland North America. There is no corresponding barrier in Europe. The tempering influence of the Pacific Ocean

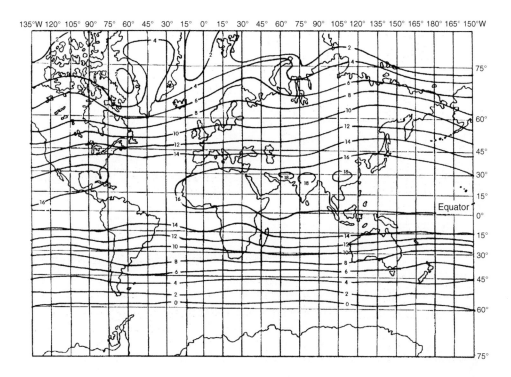

Fig. 7.2 Height (in thousands of feet) of the 0°C isotherm in July.

is therefore confined to the coastal regions of western North America, while that of the Atlantic Ocean extends far into the European continent.

The situation in Northern India sees the Himalayas effectively blocking the cold continental air to the north and preventing any really cold air penetrating over the Indian subcontinent. North America has no transverse mountain barrier, so has a completely different weather regime. The warm moist air from the Gulf of Mexico is able to penetrate northwards into Canada, and cold excursions of Arctic air can sweep unobstructed across the entire continent to the Gulf of Mexico in the opposite direction.

It can be seen from Figs 7.1 and 7.2 that the height of the 0°C isotherm shows considerable variation throughout the year in the northern hemisphere. This is a result of the seasonal temperatures of the vast landmasses. The temperature variation in the southern hemisphere does not show these excursions; in fact there is very little difference between summer and winter in the southern hemisphere from about 20°S except over South America.

7.3 Height of the tropopause

In Table 7.1 figures are given for the mean height of the polar and tropical tropopause. The polar and tropical tropopause are usually discontinuous in

Table 7.1 Mean heights of the tropopause, polar and tropical latitudes (heights in feet).

Lat.	January Polar	January Tropical	July Polar	July Tropical
70°N	29000	–	34000	–
60°N	30000	–	36000	–
40°N	35500	54000	43000	51000
20°N	–	56000	–	54000
EQ	–	56500	–	54000
20°S	–	54500	–	53500
40°S	38000	51500	34000	49000
60°S	30000	–	31000	–

the region of latitude 30° North and South (see Fig. 3.4 which shows the location of the subtropical jet stream, which marks, the thermal boundary between tropical air and polar air.) There is also a discontinuity at about 50°N (in the northern hemisphere) where we find the mean position of the polar front (transitory) jet stream(s). The equivalent situation in the southern hemisphere, the mean position of polar front, is closer to the Equator due to lower superficial landmasses.

Further to the variations shown in Table 7.1, mention should be made of the low January value ($\cong$ 28 000 ft) of the polar tropopause near Sakhalin (50°N, 140°E), whereas in July the values are high for this latitude, being some 4000 to 5000 ft higher over Siberia and also Alaska.

Chapter 8
Polar Climates

The north and south poles show contrasting areas. The southern polar region is largely an elevated plateau, and the northern polar region is an ocean. In the Antarctic, 55% of the area is above 6500 ft, and about 25% is above 10 000 ft. The highest mountains belong to the Sentinel Range and Ellsworth Mountains where the summit of the Vinson Massif (78.6°S, 85.4°W) rises to 16 900 ft.

The central Arctic, however, is covered by a thin but permanent ice pack, which extends to the continents during the northern winter. The Arctic Ocean is almost completely landlocked except for one main access point to the warmer waters of the Atlantic Ocean between Greenland and Norway.

The polar regions are shown on the 'ideal global pressure' distribution (Figs 1.2 and 1.3) as regions of general atmospheric subsidence, though the climate is not particularly anticyclonic and the winds are not necessarily easterly. The moisture content of the air is very low because of the intense cold, and horizontal thermal gradients are normally weak, with the result that energy sources do not exist for major atmospheric disturbances, which are rarely observed. The climate of both regions is very much controlled by the radiation balance and the interaction of radiation with the surface, together with its modification by atmospheric heat advection. The situation over the south polar regions is more complex and the boundary of the Antarctic is not so easily established.

8.1 Inversions

There are two main types of semi-permanent inversions in the world: the first type is found in the subtropics (see Chapter 2, section 2.7) and the second in the polar regions. While the inversions of the subtropics are normally dynamic in origin, the polar inversions are complex and are maintained not only by subsidence of warm air but also by intense surface cooling.

Temperature inversions dominate the Polar Ocean practically throughout the year, since no month averages less than about 60% inversion conditions, and in late winter the frequency reaches 100%. In winter over 80% of the inversions start at the surface and are not broken down by the invasion of occasional depressions, for the advection of warm air seems to stabilise the surface conditions. A similar phenomenon is found in the Antarctic where, with the exception of two summer months, an inversion is an ever-present normal feature over the high plateau. It is also a frequent occurrence over the rest of the continent, including the coastal regions.

8.2 Surface temperatures

Over the Polar Ocean a relatively thin layer of cold air covers the region. Air temperatures near the surface are primarily dependent on the ice surface. In summer, the prevailing melting of snow and ice holds the surface temperature close to 0°C, but positive temperatures are usually observed near the Pole in the second half of July.

Over the Arctic pack ice, the winter temperatures are found to remain constant for a considerable time. This represents a balance between the heat loss from the ice and snow and by radiation, heat conducted through the ice from the underlying water, and also the heat transported into the area by intense advection of warm air in depressions.

The minimum air temperatures in winter occur when the net radiation loss is only balanced by the transfer of heat through the ice from the underlying water, and this produces temperatures of about −40°C, or even down to −50°C if the ice is thick. The maximum temperatures in winter are a close linear function of the wind speed, because heat is transported downwards from above by the surface friction (turbulence dispersal).

The relief of the Antarctic continent, where large areas are elevated above sea level, complicates surface temperature profiles. In January, the average temperatures vary from −30°C towards the centre of the continent to about −2°C along the coasts, while in July (southern winter) the coldest areas of the plateau average −70°C and the coasts reach about −28°C.

During the winter night (April to August), the surface temperature in Antarctica and particularly over the Antarctic plateau decreases very little. This is described as a 'coreless winter phenomenon' (Wexler, 1959). When the sun sets, the temperature drops rapidly over the continent, but less so over the surrounding oceans, thus creating a strong meridional temperature gradient. This results in the formation of numerous intense depressions. The cyclonic systems which originate over the sea bring in vast quantities of warm marine air, and in this way exchange large masses of air over Antarctica above the surface inversion. This is a major factor which inhibits the rate at which the temperature drops.

8.3 Flying conditions, North polar region

In the winter, the Arctic troposphere is generally colder than that of the ICAO standard atmosphere (ISA) and temperatures in the stratosphere are also colder than standard. In the summer, the mean tropospheric temperatures are close to ISA, but stratospheric temperatures are usually higher.

Snow

Over the northern polar regions, snow falls are mainly observed from November to April; however, in the far north it is possible at any season, and near the Pole practically all precipitation is in the form of snow. The precipitation is mainly from frontal systems, and decreases towards the Pole. There is about 250 mm a year at the margins, falling to 135 mm over the Pole. Thunderstorms and hail are practically unknown, except some distance from the Pole, where they are a summer phenomenon.

The precipitation pattern over the Antarctic is not so well known however, what is known is that precipitation is mainly in the form of diamond dust. This is the description given to very small ice crystals (unbranched) that form in air supersaturated with respect to ice at temperatures below $-30°C$. Although the annual precipitation is meagre, evaporation rates are normally very small and there is a substantial ice sheet (kilometres thick) covering the continent.

Cloud

Over the North Pole, it is usually more cloudy in summer than in winter, with the exception of Greenland and the Norwegian, Barents and Kara seas where clear skies often prevail. Seasonal variation is well marked in the north and over the pack ice. The frequency of clear skies is quite high there during winter. In the north of the region and over the pack ice, overcast days are frequent in summer when low stratus cloud is extensive. Most of the sea areas are always cloudy or overcast, and often the cloud base is near the surface and tops in the order of 18 000 ft.

Visibility

In the cold Arctic air, the visibility is exceptionally good except where there is precipitation or fog. Any wind will, however, reduce visibility drastically, and can produce blizzard conditions. Advection fog is most common, but radiation fog is confined to inland Canada, Alaska and Siberia. Over the open sea and the pack ice and coastal areas, fog is infrequent in winter, but it is very common in summer, being present 50% of the time. The frequency of fogs decreases very rapidly inland during the summer. During the autumn and winter, radiation fog is most frequent inland. Over the open sea, pack ice and coasts during autumn and winter it is much less frequent than in summer. A fog peculiar to the Arctic is ice crystal fog, which forms when the air temperature is −30°C or lower. It occurs mainly over land areas, but can be observed over the pack ice. It is usually shallow with good vertical visibility, but oblique visibility will be bad.

Ice accretion on aircraft

Generally, airframe icing is less of a problem in the Arctic regions than in temperate zones. The water vapour content of air at very low temperatures is also very low. Water vapour contents of clouds in the Arctic are measured in hundredths or tenths of a g/m^3 of dry air, whereas the water content at lower latitudes reaches several g/m^3 of dry air.

It is possible to encounter clear ice, but it is infrequent and usually confined to cloud over mountain ranges. Rain ice is most likely to be a summer problem, and infrequent in the north of the region but more frequent in the south during winter and mainly over land. Rime icing is common, but severe icing is rare.

Radiation and temperature

At high latitudes the total daily solar radiation depends largely on day length which in turn varies widely with season of the year. In Table 8.1 are listed the values of day length at various latitudes.

Table 8.1 Daily duration of (possible) sunshine time on the fifth day of each month at various latitudes. (After Gavrilova, 1963).

	60	65	70	75	80	85	90
Dec	6h 43	5h 02	–	–	–	–	–
Mar	11h 44	11h 40	11h 33	11h 23	10h 50	9h 50	–
Jun	18h 49	21h 53	24h 00	24h 00	24h 00	24h 00	24h 00
Sept	12h 55	13h 07	13h 26	13h 57	15h 10	18h 15	24h 00

Table 8.1 shows that sunshine hours vary from 24 hours at 70°N and above in June to zero above the same latitude in December. The radiation climate produced with continuous darkness in winter and continuous daylight in summer is distinctly different from that found at lower latitudes.

The duration of the polar night, which is defined as the period when the solar altitude is less than 0°50′, depends on the latitude, and increases towards the Pole. Using astronomical calculations, which disregard refraction of the sun's rays and twilight, the polar night should last from 179 days at 90°N to 24 hours at the Arctic Circle. Solar refraction in practice reduces the duration of the polar night to 175 days at the Pole and zero at the Arctic Circle. In actual conditions, the polar night is shortened still further by the phenomenon of twilight, which is defined as the period when the solar altitude is between $-6°$ and $-0°\,50'$.

At the Equator, the twilight lasts for only a few minutes, whereas near the Pole it may continue for several days in succession, but it is only important for illumination, since the influx of short-wave radiation is negligible.

References

Gavrilova, M.K. (1963) *Radiation Climate of the Arctic*. Israel Program for Scientific Translation, Jerusalem, 1966.

Wexler, H. (1959) Seasonal and other temperature changes in the Antarctic atmosphere. *Quarterly Journal of the Royal Meteorological Society* **85**, 256.

Chapter 9
The Climatic Zones

Climatic zone division can be quite complex, and varies with particular disciplines. The most complex often deal with natural vegetation, or land use. However, for aviation a more simplistic approach can be adopted.

In Chapter 1, the global air circulation can be seen as a function of solar radiation (insolation) on the Earth's surface, and also of the Earth's rotation (Fig. 1.2); (see also the surface winds and their description (Fig. 1.3)). It is possible to superimpose the climatic zones on these diagrams. Climatic zones are not so well defined as the 'idealised' wind circulation presentation because of the effects of land/sea surfaces and the consequent seasonal variations in temperature. Knowledge of the climatic zones will give a general picture of the weather to be expected, although, of course, in the real world considerable modifications are seen.

The general characteristics of weather in the idealised zones of pressure and wind can be readily inferred. On account of dynamic cooling of the air by ascent, abundant cloud and precipitation forms in the temperate belts of low pressure and in the Equatorial region, while predominantly arid conditions are maintained by subsidence at the Poles and in the subtropical belts of high pressure. The seasonal movement of these systems produces transitional regions in those zones which come under the influence of one belt or another according to the season.

Thus near the solstices the tropical rains are displaced into the summer hemisphere and we are led to speak of tropical transitional regions between the equatorial low and the subtropical highs, regions which are subject to tropical rain in summer and to dry trade-wind weather in winter. Nearer the Equator there are two rainfall maxima at about the time of the equinoxes, and two rainfall minima at about the time of the solstices, that is when the sun has respectively its greatest and least declination.

The disturbances of the temperate zone extend furthest towards the Equator in winter, so that regions on the poleward fringe of the subtropical high have winter cyclonic rains and a dry summer. These elementary climatic zones have been inferred from the consideration of a uniform Earth but they are nevertheless found to correspond widely, although by no means entirely, with actual conditions. The characteristics of the zones as found on the actual Earth can now be set out in greater detail.

In this application of aviation, there are eight climatic zones, one that is unique to the northern hemisphere. Alternative climatic names are included in brackets. These zones are now described, starting at the equatorial zone and working towards the poles.

9.1 Equatorial zone (wet equatorial climate)

This zone extends about 10° either side of the Equator (20°N over Asia) and not to such a great extent over the oceans, where it is also known as the 'doldrums'. Some texts refer to this zone as the *humid tropical*. The temperature is high and so is the humidity; winds are usually light. A dry season is non-existent, but rainfall averages show two peaks associated with the passage of the sun north and south of the Equator. The precipitation is mainly in the form of showers; thunderstorms and large cumulus clouds are common. The annual total rainfall is rather dependent on topography. Remarkably uniform temperatures prevail throughout the year.

9.2 Savannah zone (trade wind littoral climate)

These zones in each hemisphere extend from the equatorial zone to little further than the tropics, about 10° to 25° North and South. They display marked variation between wet and dry seasons; the wet season is associated with the sun's latitude. As a consequence, there will be two wet seasons on the equatorial boundary of the zones, but these may merge into a single wet season further away from the Equator. Trade winds bring tropical maritime air masses from moist western sides of oceanic subtropical high-pressure cells to give narrow east-coast zones a heavy rainfall and uniformly high temperatures. Rainfall shows a strong seasonal variation; as latitude increases, so does the amount and duration of precipitation. The dry trade wind conditions are characteristic of the hemisphere winter. Temperatures are high throughout the year, and as latitude increases, so too do the diurnal and annual temperature ranges.

9.3 Arid subtropical zone (tropical desert and steppe climate)

These zones extend to about 30° North and South, and include the great (hot) deserts of the world, the Sahara, Kalahari, Arabia, Arizona, South America, and Australia. In these regions, air is subsiding on the poleward side of the Hadley cell. It is normally cloudless, and very hot in the hemisphere summer.

They also exhibit large diurnal and annual temperature excursions; the trade winds are dominant and very consistent, also there may be short rainy seasons in the bordering areas. At the low latitude boundaries, where the zone meets the savannah, rainfall is confined to the appropriate summer, whereas on the poleward margins, the rain is confined to the appropriate winter. On the west coasts bordering the oceanic subtropical high-pressure cells, subsiding tropical maritime air masses are stable and dry, but relatively cool. Foggy desert climates prevail in a narrow coastal belt (i.e. the Atacama Desert in South America). Annual temperature range is small.

9.4 Warm temperate (transitional) zone (Mediterranean climate)

This zone extends from about 30° to 45° North and South. It has a temperate (humid mesothermal) climate, and experiences wet cool winters and dry hot summers; polar maritime air masses dominate in winter with cyclonic storms and heavy rains. Tropical maritime air masses dominate in summer, and it has a moderate annual temperature range. There is a marked influence on the area bordering the sea. Similar types of zones are found in southern Australia, central Chile, the extreme south of South Africa, and parts of California. During the appropriate summer, these zones are within an arid subtropical influence. However, in the appropriate winter because of the equatorward shift of the subtropical high-pressure belt, the zones experience unsettled weather of the cool temperate zone.

9.5 Cool temperate zone (middle latitude climate)

Sometimes described as *disturbed temperate zones*. The zone extends from 35° North to about 50° North. This is the zone that sees the frequent passage of frontal (warm sector) depressions; these are interspersed by cold anticyclones or ridges. There is no dry season, and winters can be cold, or very cold in areas well away from winds directly off oceans. Gales are frequent. In the main winds are westerly; this is the zone that produces the typical weather found in England and the nearer continent.

9.6 Boreal zone (mid-latitude steppe climate)

This is a zone unique to the northern hemisphere. It becomes well developed over the northern part of North America, Scandinavia and the former USSR. The zone extends from about 40° North to 60° North. The interior of areas of mid-latitude regions that are shut off by mountains from invasions of tropical and polar maritime air are dominated by tropical continental air masses in summer and polar continental air masses in winter. This results in a much wider annual temperature range, hot summers and very cold winters, which is the dominant feature of this zone. It will be noticed that these areas do not have a mountain range barrier or large ocean between the source region of Arctic air and the landmasses to buffer the very cold air from the polar region, hence the very large annual temperature range.

9.7 Polar (tundra climate)

This zone extends north of 55° North and south of 50° South. The Arctic boundaries lie along a frontal zone where polar maritime or polar continental air meets Arctic air masses in cyclonic storms. The climate is described as humid, although very cold; there is no warm season, and no summer. The adjacent ocean water does prevent extreme winter severity.

9.8 Polar zones (perpetual frost)

These zones are the source regions of Arctic and Antarctic air masses from the great ice-caps. They extend polewards from the Arctic and Antarctic Circles. The annual temperature average is far below all other climates. There is never a monthly average above freezing. The seasonal effects are extreme and are the result of the extended daylight in the appropriate summer, and extended period of darkness during the appropriate winter. In the northern latitudes at the margins, there are areas where the surface temperature rises above freezing during the northern summer months, and some vegetation is possible, mainly

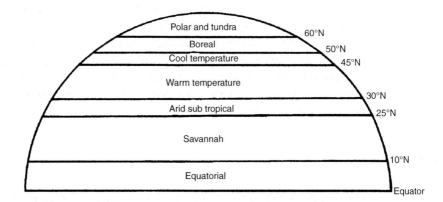

Fig. 9.1 The climatic zones for the northern hemisphere. The boreal zone is unique to the northern hemisphere; although shown as a band, it is restricted to the landmasses. In the southern hemisphere, the polar tundra climate starts at about 50°S. The southern polar zone is more extensive than its northern counterpart.

lichens and moss. However, elsewhere, the surface is covered by snow or ice and the ground is permanently frozen; in the Antarctic the high ice plateau intensifies the extremely low temperature. In northern latitudes, precipitation is mainly snow, or snow grains. In the more extensive area of the South Pole, the main precipitation is diamond dust. See Chapter 8, 'Polar Climates', and Fig. 9.1 for the broad divisions of the climatic zones.

Part 2
Route and Area Climatology

Chapter 10

Introduction and the North Atlantic

10.1 Introduction

The evaluation of aviation climatology along the air routes or over specific areas may seem somewhat formidable at first glance, but can be approached with a knowledge of the climatic zone or zones, pressure patterns and wind directions, and if there are any marked seasonal variations. However, the description of route weather will be listed *as far as possible* under the following sub-headings.

- The area
- The climatic zone or zones
- The pressure systems
- The cloud and precipitation
- The winds
- The height of the tropopause
- The height of jet streams (if applicable)
- The height of the freezing level
- The visibility
- The special or unique meteorological weather phenomena (if any).

In the following chapters, symbols are used to denote certain meteorological phenomena, see Fig. 10.1. These symbols are universally used by meteorologists.

10.2 Route climatology in the North Atlantic

See also Chapter 2, Section 2.10.

The area covered in this section includes the North Atlantic from south Greenland and Iceland in the north to the Azores and Bermuda in the south (30°N to 60°N approx.). The climatic zones will therefore be mainly 'disturbed temperate' with 'subtropical' in the extreme south. In the western portion of

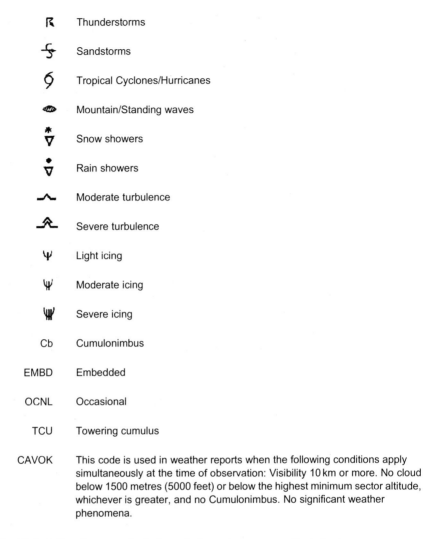

R	Thunderstorms
	Sandstorms
	Tropical Cyclones/Hurricanes
	Mountain/Standing waves
	Snow showers
	Rain showers
	Moderate turbulence
	Severe turbulence
	Light icing
	Moderate icing
	Severe icing
Cb	Cumulonimbus
EMBD	Embedded
OCNL	Occasional
TCU	Towering cumulus
CAVOK	This code is used in weather reports when the following conditions apply simultaneously at the time of observation: Visibility 10 km or more. No cloud below 1500 metres (5000 feet) or below the highest minimum sector altitude, whichever is greater, and no Cumulonimbus. No significant weather phenomena.

Fig. 10.1 A list of meteorological symbols used on aeronautical charts.

the route, commencing in NW Europe, two large pressure systems, the Icelandic low and the Azores high, usually dominate conditions.

The alternating high and low pressures of winter and summer respectively over Asia also play a part. The Icelandic depression is especially deep and extensive in winter when the temperature contrast between cold land and warm sea is greatest. It is the result of the series of frontal low-pressure systems which form on the Atlantic polar front. This front separates the cold polar maritime air (or polar continental air from North America) from the warm, moist tropical maritime air carried north under the influence of the Azores high. The position of the front varies widely, generally lying off the Atlantic coast and thence across to Western Europe. The associated depressions which form on the front move NE and are usually deepest and slowest moving while passing south of Iceland. This accounts for the fact that the lowest mean pressure is found in that region.

It follows that the North Atlantic is a region of generally unsettled weather with much thick cloud and rain and sudden changes of wind, weather, temperature and freezing level. The mean position of the main Atlantic front in winter lies approximately from the British Isles southwestwards, running north of the Azores, and thence to Bermuda and the Bahamas. In general then, the eastern Atlantic from the subtropics to about 55° North is frequently invaded by tropical maritime air. Cooling from below causes the extensive low stratus, fog and drizzle which is typical of this air mass on arrival in middle latitudes.

In the NW Atlantic the principal air mass is modified polar continental air with cumulus cloud and instability showers consequent upon the warming and addition of moisture below. Well to the south of the frontal belt, the effect of the Azores high becomes more marked and quieter settled weather with lighter winds and small amounts of cloud is general. In summer the Azores high extends its influence further northwards and the frontal systems become very much weaker and occur less often, the frequency dropping from approximately 15 per month in mid-winter to 3–5 in mid-summer. Their track is also displaced northwards.

In late summer and autumn, tropical cyclones may form in the south-west of the North Atlantic. They move west at first and then along the east coast of North America. Conditions in eastern North America are of a similar variable type, especially in the St. Lawrence area, where the depression track frequency reaches a maximum. The depressions usually alternate, however, with well-developed anticyclones and thus settled periods are more common. A further difference is that the polar air mass is usually of continental origin over the Eastern Atlantic seaboard. This air mass is unstable, but owing to the low humidity, cloud amount is not usually great. Over the sea the air soon becomes modified.

Tropical air is almost always maritime in character, originating in the SW of the North Atlantic. It is warm and moist and is cooled rapidly at the surface as it moves north along the Atlantic coast. It thus becomes stable with an inversion, and a layer of stratus or stratocumulus forms. When the air moves into the neighbourhood of Newfoundland dense and deep fog is formed and often thick drizzle. This may occur at any time of the year but is frequent in the spring and early summer when it occurs approximately one day in two.

To summarise, the outstanding feature of the region is the polar front with its associated depressions. The average position of the polar front is:

Winter   –   Florida to south-west UK
Summer   –   Newfoundland to northern Scotland

Pressure systems

The surface pressure systems affecting the area are as follows.

The Azores high

Provides tropical maritime air affecting the European coasts giving low cloud and poor visibility including advection (sea) fog, but little cloud or weather in the south of the area.

Winter – Centre pressure 1020 mb at 30°N
Summer – Centre pressure 1025 mb at 35°N

The Icelandic low

An area of average low pressure resulting from the continuous passage of depressions formed on the polar front. The depressions which form on this front move generally north-east, following the North Atlantic drift, and are usually deepest and slowest moving while passing south of Iceland. Additionally, in winter the area is surrounded by colder landmasses and therefore tends towards lower pressure.

Thus the lowest average pressure is found in this region. The name 'Icelandic low' is given to the climatic average low pressure but for any given day a surface synoptic chart will not necessarily indicate low pressure. It is situated between S. Greenland and Iceland in winter with an average pressure of 1000 mb. In summer it divides into three zones, one off Iceland, one off Greenland and the other over the Baltic Sea – average pressure 1010 mb.

Travelling polar front depressions

Frequency of occurrence will vary between about 15 per month in winter to 3–5 per month in summer.

Polar lows

See Chapter 2, Section 2.9.

North American high pressure

This will occur in winter months only, and becomes a low pressure area in the summer. It provides a polar continental outflow which reacts with warm maritime air moving north over the eastern seaboard to give heavy precipitation in the New York and Boston areas.

Siberian high

Extensions from this high-pressure area usually in the form of a Scandinavian high pressure area can affect the extreme east of the area (winter only) and bring severe wintry weather to the UK.

Tropical cyclones (hurricanes)
Can affect the extreme south-west of the area around Bermuda during August to October. They can move north-east into the Atlantic and join forces with polar front lows.

Surface winds

Mainly north-west, west or south-west, but may be easterly in the very north of the region. Can be strong to gale force in depressions and secondaries.

Upper winds

Mean equivalent headwind at 30 000 ft on the London–New York route averages 50 kt.

(1) Polar front jet streams with a mainly westerly component and speeds up to 100 kt+ throughout the year but stronger in winter than in summer due to steeper thermal gradient.
(2) Subtropical jet stream lying between 25°N and 40°N in January with maximum speeds greater than 200 kt. Further north in the summer months (≅ 40°N–45°N) but reduced strength and may not be a recognisable jet stream due mainly to a lessened thermal gradient.

Tropopause height

At 65°North   Winter 31 000 ft   Summer 33 000 ft
At 30°North   Winter 45 000 ft   Summer 47 000 ft

Height of freezing level

At 65°North   Winter 0 ft   Summer 6000 ft
At 30°North   Winter 11 000 ft   Summer 16 000 ft

Visibility

(1) *Advection fog.* Over the sea and European coastline, spring and early summer. Over Newfoundland due to interaction between the Labrador current (cold) and the Gulf Stream (warm) thoughout the year.
(2) *Radiation fog.* Inland from both European and American coastlines, mainly autumn and winter but may occur at any time.
(3) *Industrial pollution.* Near cities and industrial areas both in Europe and America, may create smog.

Chapter 11

Weather in the Arctic (North of 66°N)

North of the Arctic Circle there is continuous night in midwinter and continuous daylight in midsummer. Weather is partly maritime and partly continental in character, and near the Pole itself the latter type prevails. General weather trends can be indicated, but it should be firmly understood that there might be wide variations, as regards both place and time.

11.1 Ice

January–March

At this time the whole area is ice covered with the exception of the following.

- The southern entrance to the Davis Strait.
- The western part of the Denmark Strait and the Greenland Sea.
- The Barents Sea.

All of these are kept relatively free of ice by offshoots of the Gulf Stream and the North Atlantic drift.

August–September

By this time most of the ice has receded from the northern coasts of Siberia, Alaska and the Canadian Archipelago (except the most northern islands). The west and south-east of Greenland and the Barents Sea are clear to north of Bear Island and Novaya Zemlya. The above are, of course, mean limits and are liable to substantial variation from year to year.

11.2 Pressure distribution

Winter

At this season the main features of the pressure distribution are:

- The Icelandic low with a trough to the Barents Sea.
- The Siberian anticyclone.

- A smaller anticyclone over North America.
- The Aleutian low.

Information near the Pole is limited but available evidence suggests that it is subject to the influence of depressions alternating with anticyclonic spells, the latter being much more persistent. An almost continuous stream of depressions moves NE past Iceland towards Novaya Zemlya. Further east these depressions stagnate and fill up.

Summer

The main features are:

- A low pressure area over Iran and Afghanistan replaces the Siberian anticyclone.
- The Icelandic low has weakened.
- A shallow depression now covers much of Canada and Baffin Bay.
- Only a weak low-pressure area now exists over the Aleutians.
- Pressure is relatively high near the Pole.

During both winter and summer, depressions tend to track round the icepack from west to east, but particularly in summer they may penetrate any part of the Arctic, even moving occasionally from Canada and Alaska across the icepack to northern Russia and Siberia.

Mean sea level pressures vary from 960 to 1060 mb over the ice-cap.

11.3 Temperature

Winter

The coldest areas are the Greenland ice-cap (due to altitude and radiation) and Siberia (due to radiation). Temperatures of $-40°C$ to $-50°C$ are recorded in Siberia, and temperatures of $-30°C$ to $-40°C$ are recorded near the Pole. The lowest recorded temperature in the northern hemisphere was $-67°C$ in Siberia.

Summer

The lowest temperatures occurring are $-7°C$ to $-11°C$ over Greenland, $-4°C$ to $+2°C$ near the Pole, and $+10°C$ over Canada, Alaska and northern Siberia.

11.4 Precipitation

This averages 20 to 30 inches per year (50–80 cm) in the south and less than 5 inches (130 mm) within about 15° of the Pole, except in Greenland and the Barents Sea area. Precipitation totals show a maximum in the north in summer and a maximum in the south in winter. Thunderstorms and hail are almost unknown except rarely over northern Russia, Siberia, Scandinavia and Alaska.

11.5 Cloud

Winter

There is on the average more than 6 oktas in a broad belt from Iceland to Novaya Zemlya, Jan Mayen and Bear Island, but amounts decrease towards the Pole, where mean cover is probably 2 oktas or less. Such small amounts occur inland in Alaska, Canada and central Siberia due to the prevalent anticyclonic conditions in these areas.

Summer

Cloudiness continues high from Iceland to Novaya Zemlya and in the Aleutians. There is an increase to over 6 oktas near the Pole and a somewhat smaller increase over Siberia and northern Canada. The cloud tops of over 18 000 ft are common and orographic cloud is often deeper than might be expected. Maximum tops are theoretically at the tropopause which is at 25 000 ft in the north and 30 000 ft in the south in winter, and about 5000 ft higher in summer.

11.6 Fog

Advection fog is the commonest type. Radiation fog occurs mainly in inland areas of Alaska, Canada and Siberia. Arctic sea smoke occurs when very cold air moves over a warmer sea surface. Ice crystal fog can form over land when the air temperature is −30°C and lower. A high degree of supersaturation is possible in the Arctic without condensation, due to the absence of hygroscopic nuclei. If these are supplied by local combustion, e.g. domestic fires or aircraft engines, fog can form rapidly with temperatures of the order of −2°C to −30°C. Visibility may be very good vertically, but horizontally may be reduced to near zero. Smoke and haze are common in populated areas, where they collect under the intense surface inversions and in conditions of calm or light winds.

Winter

In winter fog occurs on 1–5 days per month over the icepack, 2–5 days per month in coastal districts and over open water, and 10–14 days per month over Iceland (radiation fog).

Summer

Fog occurs in summer on 21–29 days per month over the icepack, and 15–20 days per month in coastal districts and over open water. Fog frequency in inland areas, where it is mainly of the radiation type, is very low.

11.7 Ice accretion/freezing levels

Cumulus and cumulonimbus clouds are rare in the north and occur mainly in the south of the region. The freezing level is on the surface over most of the Arctic in winter and varies from the surface to 1000 ft in the north and from 6000 to 8000 ft in the south in summer.

Clear ice is much less frequent than other types, but the risk increases sharply when flying in the neighbourhood of high ground, e.g. over the ice-cap in Greenland.

Freezing rain is very infrequent in the north and occurs only during the summer months. Winter is the most likely season in the south.

Rime icing is by far the commonest type, and according to the USAF it will form almost every time an aircraft enters cloud or fog at sub-zero temperatures.

Altimeter errors

An aircraft flying in an atmosphere colder than ISA Standard Atmosphere would find the altimeter reading high. The figures in Table 11.1 will give some idea of the magnitude of the errors which might occur in these cold regions.

11.8 Aurora Borealis

The Aurora is an electric-magnetic phenomenon commonly called the Northern Lights and is closely associated with sunspots and geomagnetic storms. It is seldom seen in southern England and its frequency increases northwards, reaching a maximum in NW Greenland. Similar displays occur near the South Pole.

Table 11.1 Altimeter errors.

Temperature		Height		
Actual	ISA	Indicated (ft)	True	Error
−28°C	+3°	6000	5160	840
−32°C	−5°	10000	8790	1210
−52°C	−25°	20000	17740	2260
−68°C	−44°	30000	26240	3760
−77°C	−56°	40000	35920	4080

The aurora gives considerable illumination and could well illuminate the ground at night. An idea of the brightness of illumination that can be experienced can be obtained from the fact that large newsprint can be read quite easily. Radio blackouts are very common during these displays.

Chapter 12

Weather in Arctic Regions of Norway (Coastal Area)

Although the climate of northern and NW Norway is very stormy, this region is also the warmest part of the northern hemisphere within the Arctic Circle. The coast is considerably warmer than the interior, and mean temperature in January in the Lofoten Islands exceeds the world latitudinal mean by a wide margin.

Since the whole area lies within the Arctic Circle the length of day varies very greatly from season to season. This tends to create great differences between sea and land so that the winters are mild over the sea and the summers cool, whereas inland the winters are cold and the summers warm.

In the fjords, the prevailing winds are greatly influenced by local topography and quite strong winds may be experienced, *the speed being reinforced by a katabatic effect* in addition to the effect of funnelling along the valleys.

In most of the fjords some ice forms during the winter, but in the majority of cases it is confined to the heads of the fjords and shallow water. In the far north there is normally some ice at the heads of the fjords from late December or early January until about April. The west coast fjords are normally open except for the extreme tips, throughout the winter.

12.1 Effect of different air masses

Arctic air
When arctic air reaches northern Scandinavia it has been greatly modified by the passage over several hundred miles of relatively warm sea. This promotes great instability and consequently the Norwegian coast experiences extensive cumulus and cumulonimbus cloud from which there are heavy and frequent showers usually of snow between *October* and *May* and of rain, sleet or hail at other seasons.

This type of situation usually produces very low cloud in precipitation and some of the worst weather for aircraft icing, particularly over high ground near the coast. At low levels over the open sea the severity of icing is usually much less.

Polar maritime and returning polar maritime air
The majority of air masses approaching the area from the west and south-west can be included under this heading and may be stable or unstable. Although instability is not as marked as in Arctic air masses, the showers which develop are often heavy and frequent. The stable type on the other hand usually has a layer of stratus beneath the inversion, while there may be considerable orographic rain or drizzle along the hills (with snow in winter).

Tropical maritime air
This type is rare in the north and NW of Norway but when it does occur it is accompanied by dense stratus or sea fog.

Russian air masses

Winter continental air
In the far interior of Russia between *October* and *May* there is always a great mass of intensely cold continental air which has a sharp inversion below 4000 ft or even 3000 ft and usually a pall of stratus or strato-cumulus. The Russian air breaks out across northern Scandinavia on three or four occasions in an average winter, each spell lasting about three or four days, and brings intensely cold cloudy weather.

An exception occurs when the air crosses the Gulf of Bothnia. In the early part of the winter the Gulf is either completely unfrozen or thinly frozen with snow-free ice. The relatively warm sea surface destroys the inversion in the Russian air and creates instability and almost continuous snow may fall over Lapland.

Summer continental air
The source region is almost identical with that for winter continental air; it occurs only in the period *June* to *mid-September* and as far as northern Scandinavia is concerned is important only in late June, July and August. The temperature of the air is high at the surface and the lapse rate is usually fairly stable in northern Scandinavia since it has come from a warmer region. The air is dry at all levels at most times but near the surface the Gulf of Bothnia sometimes increases the humidity. The air is often involved in shallow cyclonic activity over the Baltic, and at such times its stability is greatly lessened and widespread thunderstorms may occur. The absence of night means that the thunderstorms may continue to exist for long periods.

The Arctic front

Many cyclonic waves develop on this front. Such waves most often form in the south Greenland–Iceland area, but marked local cyclogenesis takes place in the North Cape region in winter. The depressions very often pass directly across northern Scandinavia in autumn and spring. In development they vary from newly formed open waves to fully mature occluded cyclones on arrival in Norwegian waters, but the majority are still developing and move rapidly.

The Atlantic polar front

Depressions on the polar front are as frequent as those on the Arctic front. They do not affect the Barents Sea or northern Scandinavia nearly so often, however; more often they move inland across England, Scotland or southern Norway, or fill up over the north-east Atlantic before reaching northern Norway. Those that do reach northern Scandinavia nearly always approach from the south-west as old occluded centres that are often filling up.

The arctic front is sometimes involved in their circulation especially if they take a northerly track; the cyclones then often become very deep, and move far north-east across Scandinavia or the Barents Sea, Arctic air streaming south across Europe behind them. These depressions involving both major fronts are the deepest and most vigorous that develop in the northern hemisphere (excluding tropical cyclones).

Polar depressions

A small number of the cyclones reaching northern Scandinavia from the north-west appear non-frontal; these are either very old occluded systems whose occlusion is showing frontolysis, or non-frontal developments in polar air of the type sometimes observed in polar streams off Scotland or Ireland. (See Chapter 2, Section 2.9.)

Shallow summer depressions

These are shallow disturbances originating to the south or south-west over the Baltic, North Sea or southern Scandinavia. They are of the thundery type so common in central Europe at this season. Those that approach from the south often have a westward moving mass of warm Russian air on their northern flanks. This air gives the brief warm spells enjoyed by northern Scandinavia during *July* and *August*. Föhn effects sometimes give very high temperatures on the west Coast.

Chapter 13

Weather in Europe

13.1 Central Europe

Central Europe may be regarded as the transitional zone between maritime NW Europe and Continental Russia (disturbed temperate climatic zone). In winter, temperature decreases eastwards rather than northwards, and highest summer temperatures are found in the interior. Prevailing winds over most of the area are westerly, but less strong and not so constant in direction as in NW Europe.

Winters are colder than in NW Europe, and most rivers are part frozen for some period of the winter, even in eastern France. East of the Elbe temperatures below zero occur most winters. Summers are warm, very high temperatures being common in the interior. Northern Italy and the Balkan countries shares the hot summers of the Mediterranean, and remarkably high temperatures occur over eastern Scandinavia, where 37°C has been recorded on the Arctic Circle.

Precipitation is distributed fairly uniformly throughout the year, there is no true dry season, although the maximum rainfall occurs in the summer. February almost everywhere is the driest month, due to the cushioning effect of the Siberian anticyclone. However, in summer, because of the high temperatures, relative humidity is much lower than in the winter, and cloud amounts are on the average much less than in winter when the low temperatures tend to produce large amounts of cloud.

13.2 Eastern Europe

Eastern Europe consists of Russia, Finland and Poland east of the River Vistula. Here the climatic regime is continental, extremely cold in winter and extremely warm in summer; the range of temperature between winter and summer is as much as 37°C in many places. Frozen snow and ice-covered surfaces in winter are followed by floods and swamps from the melting snows of spring. In summer precipitation is of the thundery type. Nearly all European Russia has most rain in summer, but winter is far cloudier, with almost the same cloud cover as NW Europe. The clearest skies occur over the Steppes.

In winter, an extension of the Siberian high forms a 'wind-divide' along latitude 50°N. To the north, prevailing winds are from the south and west, over all northern and central Russia. Variable winds mark the wind-divide itself, and to the south winds are from the north, NE or east, very cold and dry. In summer, east to NW winds prevail over northern and central Russia, veering NE in the southern and eastern regions and continuing to the countries bordering the northern Mediterranean.

Main pressure systems (see Fig. 13.1) are:

(1)   Icelandic low;
(2)   Azores high;
(3)   seasonal high and low of Asia;
(4)   travelling depressions on the polar front.

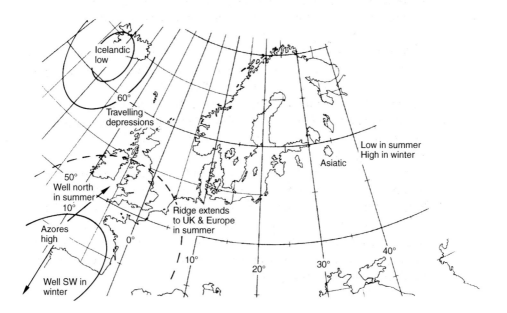

Fig. 13.1 Europe's main pressure systems.

There is no north–south mountain barrier in Europe, therefore westerly depressions can travel a long way into the continent (see Fig. 13.2). There is, however, a barrier between Europe and the Mediterranean, and this will be discussed in Chapter 14.

The air masses that affect the region vary considerably (see Fig. 13.3); they can be:

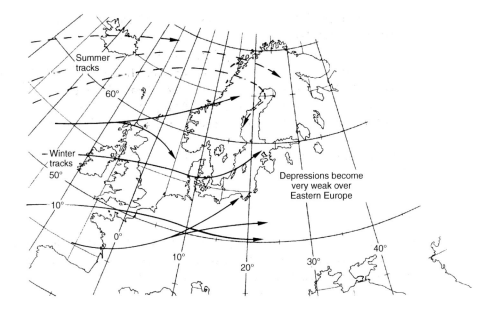

Fig. 13.2 Europe's usual tracks of depressions.

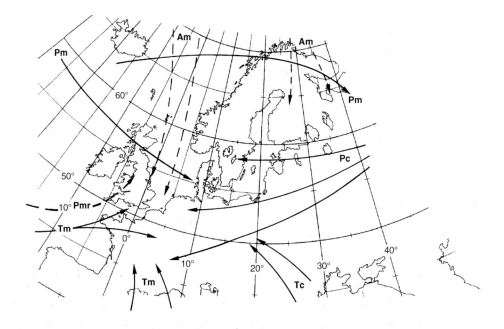

Fig. 13.3 Air masses that affect Europe.

Arctic maritime (Am)	Winter and spring
Polar continental (Pc)	Dry season
Polar maritime (Pm) and	
Tropical maritime (Tm)	Warm sector of depressions
Tropical continental (Tc)	Summer (rare)
Tropical maritime (Tm) from Mediterranean	Summer only
Returning polar maritime (Pmr)	Winter and spring

There are a number of special features that should be explained; these are as follows.

Danube low

Warm, moist air spreads northwards from central and eastern Mediterranean in winter. Gives extensive low cloud over Germany, Poland and possibly the Low Countries. Precipitation may be as snow (see Fig. 13.4).

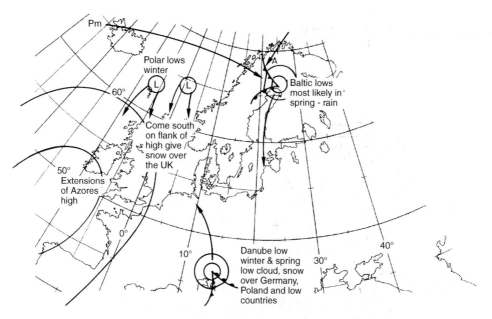

Fig. 13.4 Extension of Azores high – winter and spring thundery lows – summer. Form in Tm air (usually from Mediterranean) over Southern Europe and drift north. Outbreaks of thunderstorms often reach southern England late in evening.

Thundery lows

Form in Tm air moving north from Mediterranean over Southern Europe in summer. Outbreaks of thunderstorms often reach southern England late in the evening.

Cold front

With a depression over northern Russia, and a trailing cold front from northern Poland to France, the Alps retard the frontal progress and convergence together with orographic intensification gives a line of Cu or Cb. Waves can form small vigorous lows.

Anticyclones

These are varied but can include:

(1) Persistent extensions from the Azores high bringing tropical air in summer with clear skies, good weather and high temperatures.
(2) Persistent extensions from the Siberian high (possibly well developed over Scandinavia) giving polar continental air in winter and early spring. Producing cold dry weather with clear skies or low stratus or fog. May give cumulus and heavy showery snowfall on E coast of England (see Fig. 13.5).
(3) Transitory temporary highs between depressions can introduce cold Pm or cold Am.

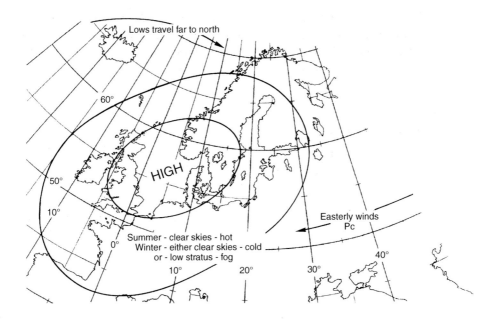

Fig. 13.5 High centred over Europe. In summer usually formed from the Azores high. In winter usually an extension of Siberian high. Can be persistent.

Winds

Generally westerly but ENE or SE from Siberian high. Westerlies increase with height (thermal component). Possible polar front jet streams around 30 000–35 000 ft, 100 kt or more in winter but generally less in summer.

Visibility

There is a high incidence of radiation fog in autumn and winter which may become widespread, dense and persistent in association with anticyclonic conditions worsened by industrial pollution. Much advection fog and low cloud (hill fog) on coasts in Tm and Pmr air. During the summer months fog is infrequent but can occur in coastal areas and over the sea.

Cloud and precipitation

All types of cloud can be experienced in the area but mean cloud amounts are quoted as 6 oktas in winter, 5 oktas in summer. Annual rainfall is 40 inches + (100 cm) in the west to less than 20 inches (51 cm) in the east. Precipitation as snow in winter particularly in the east or SE where ground may remain snow covered for long periods.

Freezing level and icing

Freezing level January 4000 ft – July 12 000 ft (for central France). In the winter often at or near the surface in the east. Heavy icing can occur due to the masses of cloud in winter. Even in summer icing risk remains high. *Rain ice can occur in winter* particularly with any easterly surface wind ahead of a warm front.

Tropopause height

January – 35 000 ft ⎫
July   – 39 000 ft ⎬ England and central France

Chapter 14
Weather in the Mediterranean

The 'Mediterranean type' of weather is described as having hot, dry summers and mild, relatively wet winters. It is interposed between the temperate *maritime type* and the arid *subtropical desert climate*, but the Mediterranean regime is transitional in a special way for it is controlled by the westerlies in winter and by the subtropical anticyclone in summer.

The Mediterranean is a land-locked sea with notable gaps (see Fig. 14.1). It is surrounded on all sides except the south-east by mountains and these play an important part in determining the meteorology of the region in several ways. They distort the air streams by causing them to flow parallel to the main ranges and in this way probably restrict the development of depressions, which remain smaller and shallower than those of the Atlantic. The mountains also determine the formation of lee depressions as will be seen, but principally the mountain barriers determine the tracks by which air masses can reach the Mediterranean. Additionally, the configuration of seas and peninsulas produces great regional variety of weather and climate.

The main gaps in the mountain barrier along the north and west coasts of the Mediterranean are:

(1) The Straits of Gibraltar.
(2) The south of France, which contains two low-level approaches to the Mediterranean: the Rhône Valley and the low ground around Toulouse. Deep cold air does, however, flow over the whole area between the Alps and the Pyrenees.
(3) The relatively lower ground which connects the Adriatic with the Danube Basin in the region of Trieste.
(4) The region around the Dardanelles and the Aegean Sea.

In addition, Atlantic air masses sometimes flow across Spain. From the south, tropical air can enter the Mediterranean freely east of Tunisia, but only with difficulty across the Atlas Mountains further west.

The winter season sets in quite suddenly in the Mediterranean as the summer eastward extension of the Azores high-pressure cell collapses. This phenomenon can be observed on barographs throughout the region, but

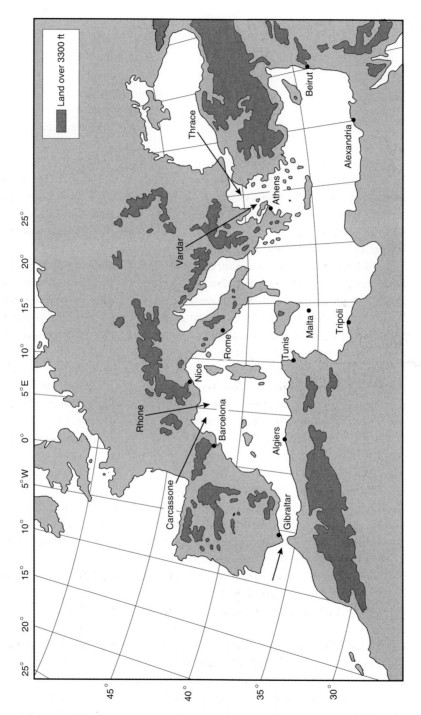

Fig. 14.1 Mediterranean Sea, general geography. It is a land-locked sea with notable gaps.

particularly in the western Mediterranean where a sudden drop in pressure occurs about 20 October and is accompanied by a marked increase in the probability of precipitation.

This change is associated with the first invasions by cold fronts, although thunderstorms and rain showers have been common since August. The pronounced winter precipitation over the Mediterranean largely results from the relatively high sea-surface temperature in winter; in January the sea temperature being about 2°C higher than the mean air temperature. Incursions of colder air into the region lead to convective instability along the cold front producing frontal and orographic rain.

Incursions of Arctic air are relatively infrequent (there being on average 6–9 invasions by Arctic air each year), but the penetration by unstable Pm air is much more common. The latter typically gives rise to cumulus development to well over 20 000 ft and is critical in the formation of Mediterranean depressions. The initiation and movement of these depressions is associated with a branch of the polar front jet stream located about 35°N.

Although some Atlantic depressions enter the western Mediterranean as surface lows, they make up only 9% of those affecting the region. Whereas 17% form in the lee of the Atlas Mountains (the so-called *Saharan depressions*, which are the most important sources of rainfall in late winter and spring), 74% of depressions develop in the western Mediterranean to the lee of the Alps and Pyrenees.

The combination of the lee effect and that of unstable surface air over the western Mediterranean explains the frequent formation of these *Genoa type depressions* (*Alpine lee depressions*) when conditionally unstable Pm air invades the region. These depressions are exceptional in that the instability of the local air in the warm sector gives unusually intense precipitation along the warm front. The unstable Pm air produces heavy showers and thunderstorms to the rear of the cold front, especially between 5° and 25° East. This warming of Pm (or Arctic) air is so characteristic as to produce air designated as Mediterranean.

The mean boundary between this Mediterranean air mass and Tc air flowing northwestwards from the Sahara is referred to as the Mediterranean front. This is a winter phenomenon. The front is located over the Mediterranean and Caspian Sea. At intervals, fresh Atlantic Pm air or cool Pc air from south-east Europe converges with warmer air masses often of North African origin, over the Mediterranean basin and initiates frontogenesis. There may be a temperature discontinuity as great as 12°–16°C across it in late winter, and this can give an indication of the severity of the weather on this front. In summer, the area lies under the influence of the Azores subtropical high-pressure cell and the frontal zone is absent.

Sahara depressions and those from the western Mediterranean move eastwards, forming a belt of low pressure associated with this frontal zone and frequently draw the warm, dust-laden Sirocco (especially in spring and autumn when Saharan air may spread into Europe). A summary of the Mediterranean winds appears later. The movement of Mediterranean depressions is greatly complicated both by relief effects and by their regeneration in the eastern Mediterranean by fresh Pc air from Russia or south-east Europe. Although many depressions travel eastwards over Asia, there is a strong tendency for low-pressure centres to move north-eastwards over the Black Sea and Balkans, especially as spring advances. Winter weather in the Mediterranean presents considerable variation. The subtropical westerly jet stream is highly mobile and may occasionally even coalesce with the southerly-displaced polar front jet stream.

The following air masses affect the Mediterranean.

(1) **Winter**
  ● Arctic (A)
  ● Polar continental (Pc)
  ● Polar maritime (Pm)
  ● Tropical continental (Tc)
  ● Tropical maritime (Tm)
(2) **Summer**
  ● Polar maritime (Pm)
  ● Tropical continental (Tc)
  ● Tropical maritime (Tm)

In addition, air which stagnates over the Mediterranean under an irregular pressure distribution is rapidly modified, and may be regarded as forming a distinct 'Mediterranean' air mass. Synoptic situations leading to the formation of such an air mass are frequent in summer but also occur in winter.

14.1 Arctic and polar continental air (A, Pc)

Arctic and polar continental air reaches the Mediterranean only in winter. It enters the Mediterranean basin by way of the Rhône Valley, the Balkan area or the Trieste Gap, and northern Italy. Initially the air is dry, cold and stable, but over the Mediterranean it is rapidly warmed and moistened, becomes quite unstable, and is characterised by vigorous development of cumulus and cumulonimbus cloud, also by showers, often with thunder and heavy icing conditions above 6000–7000 ft.

A steep lapse rate of wet-bulb potential temperature is established and the incidence of potential instability over islands and mountains can lead to thunderstorms even when showers are light and scattered over the sea. It is only on the south coast of France that these air masses are associated with fine conditions, and there they are often accompanied by the *Mistral* (see below). Under Mistral conditions vigorous development of cumulus and cumulonimbus cloud does usually occur north of latitude 42° North.

Air masses of this type do not occur in summer. The air originating over Europe reaches the Mediterranean with air temperatures higher than the sea temperature and is better regarded as tropical continental.

14.2 Polar maritime air (Pm)

Polar maritime air originates in the North Atlantic and enters the Mediterranean at all seasons; its track is over southern France or less commonly across Spain. In winter, its properties are similar to Arctic air, but less severe. Over the Mediterranean Sea it becomes unstable and showers develop. In summer, polar maritime air gives fine settled weather in the central and eastern Mediterranean.

14.3 Tropical continental air (Tc)

Tropical continental air usually originates from the Saharan region of Africa, although in summer southern Europe may also produce air masses of Tc type. On the African coast it is a hot dry air mass with a steep lapse rate, but stable because of its dryness. It is often associated with dust haze.

Over the sea it is both cooled and moistened, and as it moves northwards, it is often responsible for Sirocco conditions (see below). It is warm, sultry and sometimes accompanied by sea and coastal fog, and layers of low cloud of stratus type. At high levels this air mass contains sufficient moisture to produce rain when lifted in the circulation of a North African or western Mediterranean depression. This may be of a thundery character owing to the release of energy from convective instability.

The strong southerly winds that bring tropical continental air northwards in advance of depressions are known as *Ghibli* in the west, and *Khamsin* in the east; however, the term *Sirocco* is a generic term for most of the winds from the North African desert. The term Sirocco is quite commonly used in Crete, Malta and Cagliari, and reaches as far as Rome and Marseilles.

14.4 Tropical maritime air (Tm)

Tropical maritime air originates in the Atlantic to the south or SW of Madeira, and enters the Mediterranean as a southwesterly air stream. It is generally associated with a series of small wave depressions on a frontal zone extending along the north African coast and separating it from air of polar origin. It is not a common air mass over the eastern Mediterranean. In winter, it sometimes forms the warm sector of depressions which form near the Algerian coast and move eastwards. Also, it can form the warm sector of the Danube low pressure system (see Fig. 13.4).

14.5 Mediterranean air

Stagnant air over the Mediterranean probably tends to subside slowly and remains moderately dry. The lapse rate, however, remains near the saturated adiabatic, and although cumulus clouds and showers are unlikely to develop to any extent over the sea, they may be produced by diurnal heating over adjacent landmasses, especially over high ground.

14.6 Special phenomena

The Sirocco
The Sirocco consists of a hot current of tropical continental air, which moves from a south or south-east direction northwards across the Mediterranean. Its source is the Sahara desert and it moves northwards on the eastern side of a depression in the western Mediterranean or over north-west Africa. The extent of the area affected by this airflow depends on the synoptic situation. For example, with a small depression over southern Tunisia only Libya and the Gulf of Sidra are affected, but with a deep depression near the Balearic Islands the whole of the western Mediterranean as far north as Marseilles may be under the influence of the Sirocco.

One difficulty in discussing the Sirocco is that the local inhabitants use the term for different conditions, which vary from place to place according to the characteristics of the locality. However, if the term is used for all occasions of tropical continental air of Saharan origin, it is possible to give a consistent review of the characteristics at different places. This convention has been adopted in what follows.

The Sirocco is most frequent in spring and autumn. Although the frequency of the Sirocco is relatively low, 30 days per year in the south, when it does occur it usually persists for 2 or 3 days at a time. The characteristics of the Sirocco vary greatly from place to place mainly owing to the length of the sea track and to orographic uplift and Föhn effects.

On the coast of North Africa the tropical continental air is dry, hot and hazy. There is little cloud, but sandstorms are likely if the wind is strong. As the air passes northwards over the Mediterranean it is slightly cooled and considerably moistened so that it is recognised as a warm, sultry wind.

The stability caused by the cooling confines the moisture to the lower layers. This results in poor visibility, coastal and sea fog and layers of low stratus cloud particularly where the Sirocco is an on-shore wind. The degree of moistening and consequently the formation of fog and cloud depend on the length of time the air has been over the sea and hence on its track and speed.

At Malta, south and south-west winds have too short a sea track to give rise to low cloud or fog. A high dust haze is often observed with Sirocco from any direction although visibility at the surface is frequently not greatly reduced. Sirocco winds from the east or south-east do sometimes cause low stratus cloud or fog but neither are very common.

By the time the air reaches Sardinia or Italy (also Sicily when the air stream is south-easterly), sufficient moisture has been acquired to cause cloud, and sometimes drizzle or light rain. At Cagliari (Sardinia) the Sirocco is generally cloudy, except in summer, and visibility is less than 10 km on more than half the occasions. The Sirocco may also account in part for the very high frequency of poor visibility reported from certain coastal stations in north-west Sicily. The low cloud masses are broken up when the air stream passes over the larger islands and visibility is improved. Thus the northern coast of Sicily and the western coast of Sardinia are less likely to be affected.

One of the most widely recognised features of the Sirocco after it has crossed the Mediterranean is its oppressively high humidity. In southern Sardinia when the Sirocco occurs in July or August temperatures of 30°C may be recorded, and wet-bulb temperatures of 26°C. In the winter months, however, the midday temperature is likely to be about 18°C and the wet-bulb temperature about 13°C.

At Malta in July and August a south or south-west Sirocco wind may cause maximum temperatures exceeding 35°C, and the wet-bulb temperature occasionally reaches 26°C. More commonly the maximum temperature lies between 28° and 32°C, and the wet-bulb temperature between 22° and 24°C. With south-east Sirocco winds the maximum temperature is somewhat lower, but, as the water vapour content is higher, the wet-bulb temperature is about the same as with south or south-west winds.

The Mistral

The Mistral is a strong and sometimes violent wind that blows from the north or north-west over the south coast of France and the Gulf of Lions. It is associated with outbreaks of polar air into the Mediterranean from the north, the airflow being deflected across the relatively lower ground between the Alps and the Pyrenees by the barrier of the Alps. The onset of the Mistral usually takes place when the cold front of the last of a series of Atlantic depressions moves southwards over southern France and high pressure builds up behind it, often as a ridge of high pressure extending from the Azores anticyclone towards France.

The Mistral is usually associated with fine, cold weather and little cloud, except at the passage of the cold front, which initiates it, and of any secondary fronts which may move south in the polar air stream. The Mistral may occur at any season but is most common in winter and early spring. It blows with greatest force during the middle of the day. On one occasion in four it lasts for three consecutive days or more and has been known to continue for as long as 12 days.

The strength of the Mistral varies considerably from place to place. The strongest winds usually occur in the region of Perpignan and near the mouth of the Rhône. The strong winds continue seaward over the Gulf of Lions and, when a depression over Italy or Corsica is well developed, may continue to the African coast and the Malta Channel as a relatively narrow stream (180 to 280 km wide).

During the Mistral, upper winds are usually north-west or north up to at least 10 000 ft without decrease in speed. *Considerable turbulence can be expected in the Mistral current.*

Alpine lee depressions
Alpine lee depressions affect the Mediterranean in the winter. They form at first as fine weather depressions to the lee of the Alps with strong north-westerly winds over Europe. The Alps and Pyrenees (acting as a barrier) hold up a cold front or cold occlusion (see Fig. 14.2). The pressure generally falls in the north-west Mediterranean, particularly over the Gulf of Genoa where a shallow depression may form. The cold air mass enters the Mediterranean across the south of France unable to cross the Alps initially. The lee depression then deepens rapidly and the Mistral eventually pushes the frontal interface through the gap in the south of France. It is absorbed by the depression, which

Fig. 14.2 Cold front held up by mountains. Pressure falls south of the Alps. Fine weather in the Mediterranean at this stage.

then becomes a frontal depression and moves down through Italy, or south-wards along the Italian west coast.

There is no warm front. Part of the cold front goes round east of the Alps and is brought into the Mediterranean by the *Bora*. This adds a second cold front to the depression. Heavy rain occurs in the Po valley. The depression moves into the central Mediterranean and the cold fronts give thick cloud, heavy rain and thunder over the North African coast and the Atlas Mountains.

The depression then moves towards Cyprus, drawing in more cold air from Greece, generating another cold front. The cold fronts give sandstorms on the North African coast (see Figs 14.3, 14.4 and 14.5). The Cyprus low usually fills up *in situ* but occasionally one passes across Iraq, can penetrate the Persian Gulf, and sometimes reaches North-west India. (See Figs 16. 1 and 17.1.)

Fig. 14.3 The Mistral breaks through and a cold front is pushed into the Mediterranean. No warm front weather – the Sirocco is too dry. Centre of low moves south – heavy rain in the Po valley.

Atlantic depressions sometimes break through the Gibraltar gap or the Carcassonne gap in the Mediterranean (see Fig. 14.6). Again there is no warm front since the warm sector air is drawn from the deserts of N Africa and is very dry. The cold fronts weaken after crossing Spain but are rejuvenated by the *Mistral* and are quite active as they move towards Malta. They then weaken in the central Mediterranean as the polar air is warmed by the sea. It is again re-activated by the cold *Vardarac* and the low becomes a Cyprus low as before (see Fig. 14.7).

Regional winds are also related to the meteorological and topographic

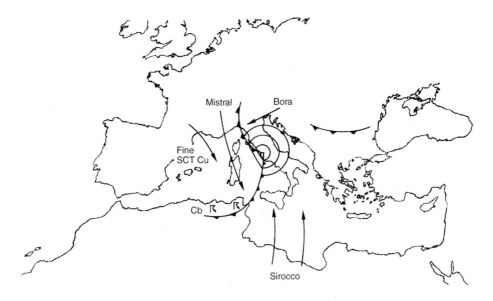

Fig. 14.4 Following from Fig. 14.3, the depression deepens and moves SSE – heavy rain. The bora generates a second cold front. Cbs over the Atlas Mountains.

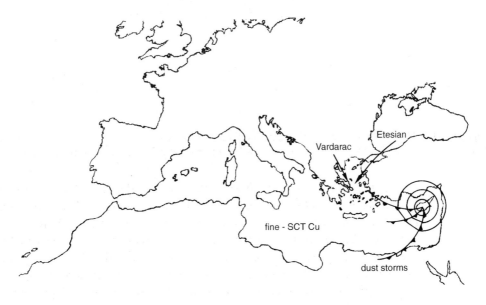

Fig. 14.5 The depression finishes up as a 'Cyprus low'. Ns, EMBD Cb near the centre. Cu and Cb along the cold fronts.

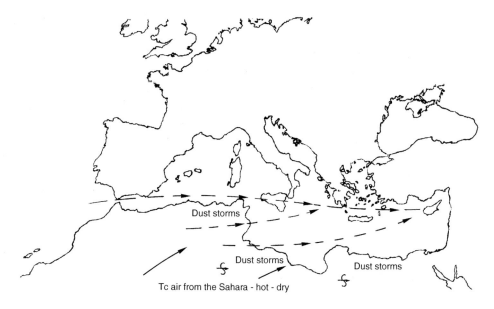

Fig. 14.6 Tracks of depressions from the Atlantic through the Gibraltar gap and Atlas lee depressions. They are weak in the west, dried out over Spain, squally winds raise dust. They become more active in the east – centres of lows and cold fronts likely to be very active if they track to the north.

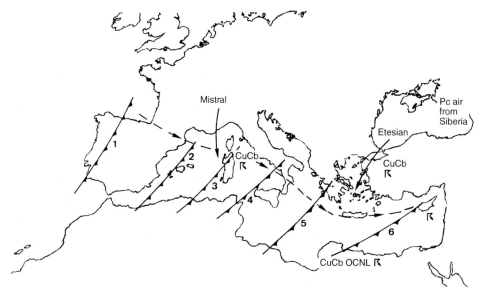

Fig. 14.7 1. Cold front active on arrival from the Atlantic. 2. Weak, dried out over Spain. 3. Active, rejuvenated, warm sea acts on fresh Pm air. 4. Weak, Pm air now warmed by the sea. 5. Active, Pc air now in circulation of Cyprus low.

factors. The familiar cold, northerly winds of the Gulf of Lions (the Mistral), which are associated with northerly Pm airflows, are best developed when a depression is forming in the Gulf of Genoa east of a high-pressure ridge from the Azores anticyclone. Katabatic and funnelling effects strengthen the flow in the Rhône valley and similar localities so that violent winds are sometimes recorded. The Mistral may last for several days until the outbreak of polar or continental air ceases.

Similar winds may occur in the northern Adriatic (the Bora) and the northern Aegean Sea when polar air flows southwards in the rear of an eastward-moving depression and is forced over the mountains. The generally wet, windy and mild winter season in the Mediterranean is succeeded by a long indecisive *spring lasting from March to May*, with many false starts of summer weather.

14.7 Khamsin or Ghibli depressions

These depressions form over the desert regions of North Africa as wave depressions on a frontal interface separating air which has been over Africa for some time and has been considerably warmed, from polar air which has recently crossed the Mediterranean and is much cooler. As the warm air mass is very dry the cloud associated with these depressions is not extensive. The warm front, if it can be traced, is diffuse, but the cold front is usually clearly defined.

The depressions often form south of the Gulf of Sidra, but they may form much further west; indeed the origin of some of the depressions, which are of importance over Tripoli, can be traced back to the passage of a cold front over the Atlas Mountains. Depressions forming well to the west frequently turn north over Tunisia and pass on over Italy and the Adriatic. Khamsin depressions which first appear over Libya usually move eastwards towards Egypt, but sometimes they cross the Libyan coast and move out over the Mediterranean. Less vigorous depressions of the same type also form nearer to Egypt moving eastwards across the Nile Valley.

The Khamsin (or Ghibli) depressions are preceded by strong southerly winds which cause widespread sandstorms as well as excessively high temperatures. The passage of the cold front is marked by a rapid change of wind to north-westerly. In some cases the north-westerly wind is sufficiently strong to maintain rising sand for a while. More often, however, visibility shows an almost immediate improvement, although dust haze may remain for some time.

Ahead of a Khamsin depression the sky is often cloudless, or cloud is at medium and high levels. Once the cold front has passed, the increased relative humidity of the north-westerly air stream causes the formation of cumulus or cumulonimbus type clouds, and showers may occur near the coast. The cloud is seldom low enough to be a hazard to flying. Subsidence in the cooler air stream is fairly rapid and broken stratocumulus cloud is common in the morning immediately following the passage of a cold front of this type.

Khamsin depressions are most common in the months from February to June, being almost as frequent in spring as depressions moving east along the Mediterranean. Two or three such depressions a month may be expected in March and April but fewer at other seasons.

14.8 Sandstorms, duststorms and rising sand

Along the North African coast, including the area as far east as Cairo, the reduction of visibility by sand and dust raised by the wind is one of the most important meteorological features. There are two distinct sets of conditions under which visibility may be reduced. The first occurs when strong winds exist over a wide area of desert usually associated with a Khamsin depression. Dust is then raised several thousand feet and carried long distances. The second type occurs when the wind freshens temporarily or locally over areas of loose sand, the sand is not carried far and visibility is only reduced in the vicinity where the sand is raised. Therefore a distinction is made between 'rising sand' and the true 'sandstorm' of the Khamsin type. Meteorological reports often use the term sandstorm to mean either dust or sand. Sand requires a wind speed of at least 16 kt to lift it because of the particulate size. It can reach heights of about 200 ft in a strong wind. Dust, however, can be lifted to far greater heights, even to the tropopause.

The extensive type of sandstorm is usually associated with a period of strong southerly winds before the passage of the cold front of a Khamsin depression. Such storms may, however, also occur with south-west winds in advance of the cold front of a vigorous depression over the Mediterranean. When associated with a Khamsin depression, bad visibility is persistent, although it is usually somewhat better at night than by day.

Visibility may be reduced to less than 50 m and poor or bad visibility may last for periods up to three days. The area affected may exceed 200 km in extent and usually moves east with the depression which causes it. The dust-laden air can extend above 10 000 ft. Such duststorms are most frequent in spring when the Khamsin depressions are most active. *They do not occur in summer.*

On the other hand, reduction in visibility by rising sand is of a local and temporary character. It occurs only within 3–4 h of local noon when wind and convection are strongest, and its occurrence and severity are largely influenced by the nature of the local ground surface. The presence of crop monoculture or other areas of large cultivation reduces its incidence.

Visibility in rising sand of this character may fall to 200–500 m but it is usually variable and it is often possible for an aircraft to land at an affected airfield during a temporary lull. The sand does not usually rise more than about 200 ft. Visibility may be reduced by rising sand of this character at any time of the year but most frequently in spring and summer. Any synoptic situation giving rise to a fresh or strong wind may cause it.

14.9 Summary of local winds in the Mediterranean

Vendevale   Strong SW to west wind in the Straits of Gibraltar blowing ahead of a cold front crossing Spain. September to March. Very squally with much low cloud.

Levanter   Easterly wind blowing over the Rock of Gibraltar. Usually moderate but may reach gale force. Any time of year but mainly *March and July to October*. Air very moist, hence low stratus covering the rock and drizzle. *Extremely turbulent on the approach to, and over the runway*. Turbulence can extend a few kilometres or more downwind and up to 5000 ft.

Sirocco   General name for all the winds blowing from North Africa. The local names are given on the map (Fig. 14.8). Very hot, dry and dusty. They blow ahead of depressions, hence mainly in winter. Most frequent in *September, October and March to May*. Visibility can be 100 m on the African coast, and dust can be carried well across the Mediterranean, e.g. to Malta. If the wind is light it picks up a lot of moisture from the sea and can then give low stratus and drizzle, e.g. at Malta where it is hot and humid.

Mistral   Strong northerly winds blowing down the Rhône valley, also strong WNW winds through the Carcassonne gap. Often gale force, sometimes over 80 kt. Can last for up to five days.

Bora   Similar to the Mistral, but blows through the Trieste Gap. Onset often sudden with no warning. Can last up to 30 days.

Marin   Southerly winds in the Gulf of Lions blowing ahead of a depression. Very moist, hence much low stratus and drizzle. Very warm and humid.

Vardarac   Similar to the Bora. Northerly wind blowing through the Vardar gap. Strong and squally. Winter phenomenon.

Etesian (or Meltimil)   Steady north-easterly winds from the Black Sea in summer, becoming northerly over the Aegean.

Gregale   Strong north-easterly wind in the central Mediterranean. Mainly winter. Brings very low cloud, rain and poor visibility over the airfields, with gusty winds. Usually lasts 2–5 days. See Fig. 14.8 for location of local winds.

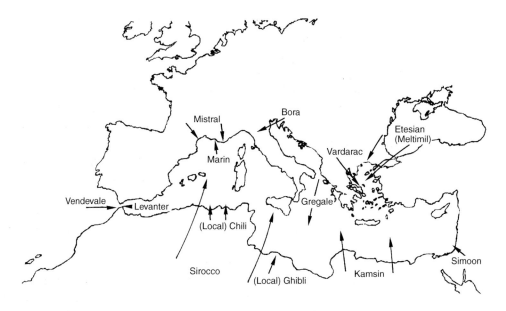

Fig. 14.8 Local winds of the Mediterranean.

Upper winds

January – westerly. Northern flank of the subtropical high with the subtropical jet stream over the North African coastline or Cyprus in the eastern Mediterranean. Generally 260°T/80–120 kt max, November to March. *The subtropical jet stream does not affect the western Mediterranean.* Polar front jet streams may invade the whole area occasionally but are an irregular feature.

July – westerly. No jet stream evident, only strong winds, i.e. less than 80 kt, aloft in the E. Mediterranean. Western Mediterranean maximum average speeds of 40 kt.

Visibility

Western Mediterranean
General. Visibility over the western Mediterranean is usually good, exceeding 15 km. Such restrictions that do occur are associated with well-defined synoptic situations.

(1)   Depressions in the Gulf of Genoa or Tyrrhenian Sea. Visibility may be severely restricted by heavy rain for short periods, and also more persistently by hill fog on the high ground in west Sardinia and north-east Tunisia.

(2) Sirocco-type situations. Any synoptic situation giving south or south-east winds of tropical continental origin with a long sea track reduces visibility over wide areas. The decrease in visibility, which is due to an increase in moisture in the lower layers of an initially dusty and hazy air mass, does not generally cause visibility less than 2 to 6 km, but occasionally when winds are light fog patches may result. A dust haze often exists to considerable heights. Visibility in the lower layers tends to decrease as the air moves northwards over the cooler sea. A variant of the Sirocco type, which brings the worst visibilities in this region, occurs with a depression over north Libya or the Gulf of Sidra. An easterly current of Tc air then affects the area. The rare occasions when fog occurs at Tunis are almost invariably associated with this type of situation. Low stratus cloud may also occur at night on the coast and low ground of eastern Tunisia.

(3) Slack-pressure distributions. When air is stagnant over the Mediterranean some local morning mist or fog patches may occur over land areas in Sardinia and Tripoli; this occurs more particularly in the summer months. Radiation fog may occur over the low-lying ground north-west of Cagliari.

(4) Sandstorms. Strong winds over Tripoli may raise dust and sand and give rise to sandstorms. The reduction of visibility depends largely on the nature of the ground surface and of the vegetation. Although strong winds from any direction may raise dust, sandstorms are most common with the strong southerly winds in advance of North African depressions.

Central and eastern Mediterranean

There is the usual deterioration of surface visibility with precipitation from any cold front. Possible coastal and sea fog over the European coastline with the Sirocco due to advection. Fog and low stratus is possible during June to August over Egypt in the early morning, with a light northerly wind usually clears 0800–0900 local time. Poor visibility in the Adriatic associated with the Bora.

Sandstorms occur over North Africa with any surface wind over 16 kt and are commonly associated with depressions with strong southerly winds. Reduction of visibility depends upon the ground surface and type of vegetation but can be down to 10 to 20 m. Vertical development is variable but may reach between 5000 and 10 000 ft. Resultant effects may be found in southern Italy and southern Spain, where visibility can be greatly reduced in dust.

The occurrence of radiation fog in the Nile Delta is most frequent between May and September, the incidence being highest at Cairo International which is 500 ft above sea level. Here there is an average of seven occasions per month. This fog occurs late at night and clears quickly in the morning.

The most favourable situation for radiation fog arises when warm air from the land at the eastern end of the Mediterranean (including Iraq) moves westwards over the Mediterranean and is returning over Egypt as a light north-westerly wind. Radiation fog also occurs during winter but its frequency is much less than in summer. The position of Cairo International makes it more vulnerable to rising sand and dust than most other airfields in the Delta. In summer serious reduction of visibility on account of sand and dust is rare and seldom prolonged. Sandstorms of short duration sometimes occur when a shallow layer of cool air from the Mediterranean moves in over the desert during the day.

In winter, however, sandstorms occur on an average six times a month, being most severe in the deep cold air currents affecting the area in January to March. Normally, airfields on the eastern side of the cultivated areas remain approachable longer than Cairo International, although the more severe sandstorms often reduce visibility to less than 500 m throughout Lower Egypt.

At Cairo International winds of 16 kt or more from between west and south give sandstorms in which visibility falls below 500 m for several hours. Winds of similar strength from between the north-west and the north-east cause only temporary sand flurries in which visibility falls to about 2000 m in a shallow layer. Air-to-ground visibility is then not seriously affected.

The formation of low stratus cloud at night in the Nile Delta is a common occurrence in summer. Its base is frequently 600 ft or lower, and at Cairo International it may be on the surface. It tends to form on successive mornings for a period of 3–5 days, but usually disperses within 3 h of sunrise.

In winter such low stratus cloud is rare, but stratocumulus cloud based at about 1000 ft or above may occur in the mornings.

Freezing level

Winter   4000–8000 ft
Summer   14 000–16 000 ft

Icing risk is high in cold frontal conditions and care is required when approaching the Alps from the south over northern Italy in winter when the 0°C isotherm is frequently down to ground level.

Tropopause height

Winter   40 000 ft
Summer   44 000 ft

Chapter 15
Weather in Africa

Africa is a vast continent, it extends from 30°N to 35°S. There are widely differing climatic zones, but two areas are well represented, the two subtropical high pressure regions, the Sahara Desert in the north and the Kalahari Desert in the south (see Fig. 15.1).

Fig. 15.1 Africa – Topography. The two desert areas indicate the locations of the subtropical high-pressure regions.

In these notes an attempt has been made to separate the African weather into generally west and east, with the northern (Mediterranean) coast having been considered in the previous chapter. The climatic zones are 'subtropical high pressure' in the north and south of the continent, with the 'Equatorial trough' in a mobile position, seasonally variable. The Sahara Desert and the Kalahari Desert affect the climate and there are no large mountain barriers to inhibit air mass movement.

The climatic regions extend from the influence of the disturbed westerlies of the northern hemisphere in the northern winter through the arid subtropical climate of the lower latitudes into the equatorial rain belt. Thence through the region controlled by the subtropical anticyclone of the southern hemisphere to South-west Africa, where the variable westerlies of the temperate latitudes are found in the southern winter.

15.1 Movement of the ITCZ

The seasonal variation over much of the central portion of the route is due to the movement of the ITCZ, which swings from approximately 15° to 20° North in the northern summer to similar latitudes in the southern hemisphere in the southern summer. (Compare corresponding movement in West Africa.) See Fig. 15.2.

Figure 15.2 shows the mean position of the ITCZ from its most northern excursion in July to the southern limit in January. There is a distinction shown in Fig. 15.2 between the oceanic and continental ITCZ. Although the ITCZ is formed by convergence over both land and sea, there is considerable local modification, and diurnal variation over the land. It looks at first sight as if there is a considerable excursion across the landmass between the two extremes. Usually, the ITCZ is well in evidence on its northern movement across the desert areas, and indeed also in East Africa, bringing the long rains. It is also well marked over the oceanic areas. Notice that the excursions are only about 5 or 6 degrees of latitude off the West African coast (a cold water coast), whereas they are some 25 degrees at the east coast (a warm water coast), and even more over the central land mass.

The position of the ITCZ is not so easily identified over central and southern Africa, particularly in January. The southern hemisphere enters its warmer phase in January, and the landmass south of the Equator experiences low pressure. There are considerable cloud developments, and it is difficult to distinguish a marked convergence line. The positions of the ITCZ indicated in Fig. 15.2 over central and southern Africa can only be taken as a general guide.

Conditions south of the Equator are in summer influenced also by the frontal zone which forms at that season, parallel to the coast of East Africa, between the moister north-easterlies of the winter monsoon of the Indian Ocean and the

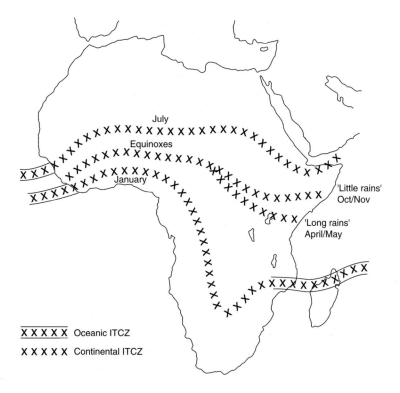

Fig. 15.2 The mean position of the ITCZ across Africa throughout the year.

drier northerly winds inland. It must be borne in mind, however, that the topography of the country varies considerably and thus there is pronounced local variation on account of influences such as altitude, lakes, etc.

15.2 The North-African coast to the Red Sea

October to April

The depressions over the eastern Mediterranean and North Africa affect the weather of the Red Sea, particularly in the north. Fronts associated with these depressions tend to become stationary near 20° North, and rarely reach beyond 15° North. Surges of moister air from the Indian Ocean affect the southern half of the Red Sea and may extend further north. Cloud and rainfall amounts are generally small. Some showers and perhaps one or two thunderstorms occur in the north at cold fronts. Rainfall is greatest along the African coast between about 15° and 20° North, but it is confined largely to the coast, and at Asmara as well as further west it is negligible except for a few showers early and late in the season.

In the southern Red Sea there is some early morning low cloud along the

African coast and near Aden. Also showers are likely near Aden and Djibouti when the monsoon is unusually deep and from a more easterly direction clear of land. Khamsins occur in the north, and duststorms associated with cold fronts occur further south-east of Khartoum and in the central region of the Red Sea. There is also some hill fog on high ground facing the north-east monsoon, chiefly at night. See Fig. 15.3 – Africa July climate, and Fig. 15.4 – Africa July rainfall.

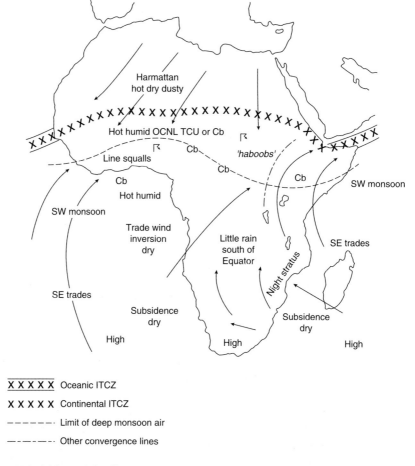

X X X X X Oceanic ITCZ

X X X X X Continental ITCZ

– – – – – –· Limit of deep monsoon air

– – · – – –· Other convergence lines

Fig. 15.3 Africa – July climate.

May to September

The low over Asia and its topography controls the circulation in the lower levels over the Red Sea, and a dry northerly air stream (west near Aden) prevails. The West African south-west monsoon extends almost to the south of the Red Sea, and on occasions advances across it.

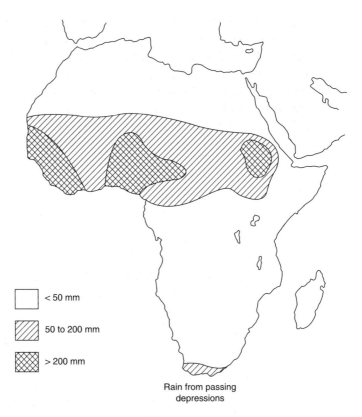

Fig. 15.4 Africa – July rainfall. Most rain occurs in the region 270 to 800 km south of the Continental ITCZ. Cbs and thunderstorms, often violent. Line squalls at the west coast. Rain is also heavy around the convergence line over Ethiopia.

Cloud and rainfall are negligible in the north. Further south, near Port Sudan, there is some daytime Cu that on occasions persists throughout the night, but rainfall is still negligible.

Over the southern Red Sea there is rather more cloud, but it is mainly Ac and high cloud. Inland, however, over high ground Cb develops in the afternoon and thunderstorms occur. On occasions these storms drift towards the sea. Over high ground just south of Asmara there are more than 15 days each month on average with thunderstorms from June to September, but about 10 days or less for the whole season is usual for the African side of the southern Red Sea.

Khamsins still occur early in the season, but in Egypt reductions in visibility from dust occur mainly about midday. Further south, *Haboobs* and pressure gradient type duststorms are most frequent in the region south of Port Sudan and east of Khartoum. In a strong pressure gradient dust may be carried into the central Red Sea. Just south of Port Sudan severe duststorms occur as a

result of a funnelling effect through the Thokar Gap. They may last intermittently for a week at a time.

In the southern Red Sea area, duststorms are almost confined to the period
April–October, with a maximum frequency in July–August. Line squalls affect
the region near Khartoum, mostly in August–September. They generally form
along the Ethiopian foothills near Kassala or further north along the west side
of the Red Sea hills and move west-south-west. The former affect Khartoum
and are generally more severe than those further north. See Fig. 15.5 – Africa
January climate, and Fig. 15.6 – Africa January rainfall.

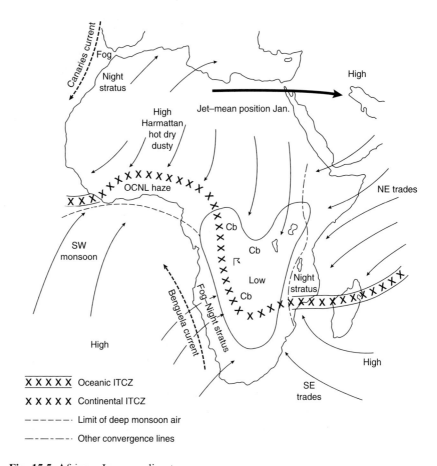

Fig. 15.5 Africa – January climate.

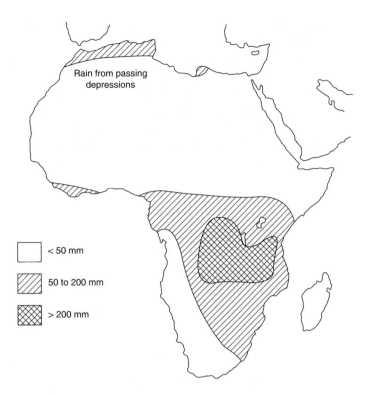

Fig. 15.6 Africa – January rainfall. Heavy rainfall south of Equator due to convergence at ITCZ (which oscillates over a very wide area) and at the other two convergence lines, amplified by orographic effects. Most rainfall is from Cb and heavy thunderstorms.

15.3 Khartoum to tropical East Africa

Mid-December to February

In the north, the sky is cloudless much of the time, and the only hazard is the occasional winter type duststorms that occur near Khartoum, mainly in the second half of this period.

Near the Equator there is some early morning low cloud and daytime Cu, with occasional showers south of about 6° North. Isolated thunderstorms form in mountainous regions, but near Entebbe Sc, Cu and Cb develop regularly at night.

March to June

Cloud spreads north in step with the ITCZ. By day Cu type clouds are common, and if development is general they merge into a cloud sheet with embedded Cb, giving widespread rain during the early part of the night.

The extreme north has a few thunderstorms, mainly in May–June, and the number increases towards the south. Most of them still occur near Lake Victoria (20 days of thunder each month just to the north of the lake), where many are nocturnal and severe in April to May.

Towards the end of this season thunderstorms are probably as frequent over the Ethiopian highlands just to the east of the route to Nairobi. East of Lake Victoria the number of thunderstorms falls markedly, and although at Nairobi, March–April is the main period the frequency is small compared with Entebbe. In these two months it is also greater near Dar-es-Salaam (7–10 days of thunder each month) where nocturnal thunderstorms are liable to occur with the onset of the south-east monsoon in April.

Winter type duststorms occur near Khartoum in the first half of this season, and summer ones (Haboobs and pressure gradient storms) can occur as early as April. Haze is also fairly frequent during the day. Low cloud is liable to form during the night near the Equator, and to a lesser extent mist or fog can also occur.

July to September

In the north, daytime Cu and Cb with showers and thunderstorms are a regular occurrence, and they are especially well developed about 100–350 km south of the ITCZ. The thunderstorms may persist well into the night and in the early morning spread out as sheets of medium cloud which disperse during the day. The frequency of thunder is least in the north (7 days each month at Khartoum). Towards Nairobi the main area for thunderstorms is around 9° North near the Ethiopian highlands, and close to the Equator the number falls off rapidly. En route to Entebbe the worst area is still north of Lake Victoria (20 days of thunder each month) and again many thunderstorms near the lake are nocturnal. Beyond Nairobi towards Dar-es-Salaam the number is negligible.

Summer type duststorms occur in the north. Some of these are caused by line squalls which originate in the Ethiopian foothills and move west. Increasing amounts of broken medium and high cloud are present ahead of the disturbance line, which may be seen approaching by the usual line of Cbs and dark roll cloud. Behind the squall line, thick medium cloud lasts for several hours with intermittent rain. A second, less active disturbance may follow at a distance of about 300–350 km. Early morning low cloud is not uncommon near the Equator, and on occasions there is some mist or fog.

October to mid-December

Flying conditions in the north improve, but the equatorial region experiences a second wet season ('short rains') not unlike the 'long rains' from *March to May*, but as a rule the storms are not so severe.

Near Khartoum thunderstorms die out during the first half of this season. The number of storms is higher to the south, but decreases as the season advances, except close to Lake Victoria where again many are nocturnal and thunder is heard on about 20 days a month. East of Lake Victoria the frequency decreases markedly as in other seasons, but between Nairobi and Dar-es-Salaam there is an increase during the second half of the season, and Dar-es-Salaam has five days of thunder in December. In mountainous areas the number is higher.

15.4 Tropical E Africa to South Africa

Mid-October to mid-April

Medium clouds are common at sunrise, possibly with local light rain. The clouds usually break within a few hours, and during the morning Cu forms. Development is sometimes small, but on other occasions there are large Cu or Cb with showers and thunderstorms, squalls and occasional hail. The thunderstorms may persist through the night.

With rising pressure near the east coast of South Africa, SE winds bring ragged low cloud ('*guti*' in Zimbabwe) and large amounts of medium and high cloud from the east. Drizzle from the low cloud may be augmented by bursts of rain from higher levels, which may be prolonged over the Transvaal. Line squalls are liable to occur at the onset and sometimes ahead of the SE winds. The bank of Cu/Cb may be very long, but rarely more than 55–90 km wide.

Near the ITCZ there is a great deal of medium cloud and rain. At lower levels there are layers of Sc and ragged low cloud, which build up rapidly into Cu/Cb whenever the sun breaks through. Thunderstorms frequently persist through the night in the Zambezi valley.

The frequency of thunder is greatest near the large lakes, where it reaches, and in some places exceeds, 20 days a month in January–February. Further south the number decreases, and in Zimbabwe it varies between 10 and 15 days a month from November to February. Beyond about 22° South storms are both most severe and numerous near Johannesburg. Very occasionally cyclones occur in the Mozambique Channel. They can occur as early as November, but are most frequent from January to April.

Late in October or in November tornadoes sometimes occur near Johannesburg. Further south towards Cape Town the weather is generally fair with fresh south-east to south-west winds (strongest in December and January). Occasionally north-west winds bring cloud and showers. In March and April there may be a thundery spell with north to north-east winds. Low cloud may drift in off the sea by night in the extreme south.

Mid-April to mid-October

An anticyclone is often centred over or near the Transvaal, and cloud amounts are almost nil except for fair weather Cu over Malawi. On occasion a trough crosses the Republic of South Africa from the south-west, to bring isolated thunderstorms over the Transvaal and Zimbabwe in the evening.

With the approach of the anticyclone, winds are south-east, and ragged low cloud spreads westwards across Zimbabwe (the 'guti' mentioned above) and occasionally the Transvaal. It does not normally extend further north than southern Malawi, but when the moist air is deep it may extend over Zambia. If it is unusually cold, sleet or snow occurs over the Transvaal. Thunderstorms are infrequent, and occur mainly at the beginning and end of the season.

15.5 Low-level jet stream: from (Findlater 1969)

It has been demonstrated that high-energy flow, in the form of low-level southerly jet streams, has been reported over Kenya, being part of a much more extensive current of air which flows rapidly around the western half of the Indian Ocean during the northern summer. See Fig. 15.7.

The high speeds are associated with the concentration of the cross-equatorial airflow into the zone from longitude 38° East to about 55° East instead of being rather evenly distributed from 40° to 60° or 80° East as illustrated in many climatalogical atlases and charts of mean flow during the season. The high-speed current is shown to flow intermittently from the vicinity of Mauritius through Madagascar, Kenya, eastern Ethiopia, Somalia and thence across the Indian Ocean to the west coast of India and beyond. The stream is occasionally reinforced by northward flow through the Mozambique Channel.

The high-speed air current, or system of low-level jet streams, is closely associated with the ITCZ over the Arabian Sea and western India. Variations in the strength of the jet stream over Kenya during a two-month period was related to the rainfall which western India received from the south-west monsoon.

Although the jet stream is not a continuous feature, the strong flow is noticeable on about half of the days during the season of the south and south-west monsoon, and when it accelerates to peak speeds it frequently does so in narrow cores elongated along the wind direction and displaying dimensions which do not vary much from one example to another. It is apparent that the core of the jet stream, as defined by the 50 kt limit, is usually 180 to 350 km wide, 550 to 1000 km in length and, from a study of vertical wind profiles, about 3000 to 3500 ft in depth. Many other analyses confirm the relative constancy of these values.

A very extensive low-level jet stream system, such as shown in Fig. 15.7 where there are sufficient reports to attempt a detailed analysis, shows that it is

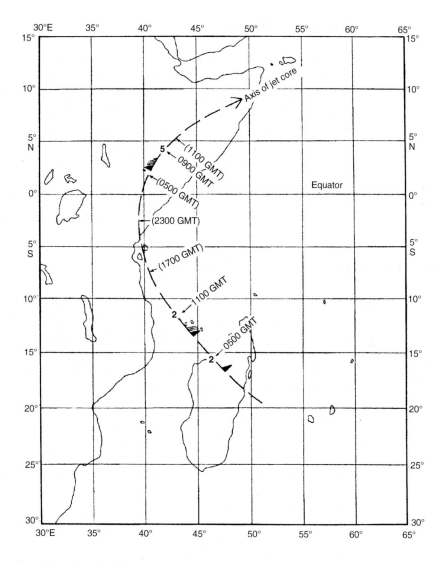

Fig. 15.7 A low-level jet stream observed over East Africa during two days in August (13–14, 1966). The height of the jet core in thousands of feet is shown at the point of the wind arrow (after Findlater 1969).

made up of a series of segments of about the dimensions noted above. Because of the general paucity of pilot balloon data at any one synoptic hour and the smallness and transient nature of the individual jet cores it was seldom possible to track an area of maximum speed downstream from one chart to another. Nevertheless, one example of a jet core which affected three stations in turn is shown in Fig. 15.7.

The jet was first located at Majunga at 0500 GMT on 13 August 1966 (Findlater 1969), when the core was at 2000 ft with a speed of 50 kt. Six hours later the core was over Moroni where 81 kt was recorded at 2000 ft, and twenty-two hours later, at 0900 GMT it passed over Mandera where 95 kt was measured at 5000 ft. (Note that the track of the jet core is passing over higher terrain from sea level to over 4000 ft above mean sea level over Kenya, and over 6000 ft in Ethiopia). The core moved at a speed of about 50 kt along the streamline, but it is likely that this speed was somewhat less in the south and more in the north where the extreme speed was recorded.

15.6 West Africa

The weather in West Africa is greatly influenced by the Equatorial trough and the ITCZ. In January and February the ITCZ is in its most southerly position, normally just inland from the West African coast but occasionally further south. In March it begins to move north and reaches beyond 20° North in July and August. Surges of the south-west monsoon occasionally extend over the Hoggar Massif, southern Algeria.

The rainfall of West Africa depends upon the interaction of three major air masses. Of these, two are normally in contact with the ground, the tropical maritime and the tropical continental, the two meeting along the ITCZ. The former is warm, moist air, which over West Africa blows inland in a general south-west to north-east direction from the ocean. It has a very high relative humidity: Warri, in the Niger delta, and lying under this air mass throughout the year, shows mean monthly values of relative humidity between 95 and 99% at 0600 hours, and between 65 and 82% at 1200 hours. The latter air mass is warm and dry, blowing in the opposite direction, from the desert, with a very low relative humidity. Sokoto, for the five months December to April inclusive, shows mean monthly values between 23 and 38% at 0600 hours, and between 10 and 17% at 1200 hours. The third air mass is the cool equatorial easterlies of higher altitude, and this will be mentioned later in connection with the West African line squall phenomenon.

The tropical continental air overrides the tropical maritime air to give the latter a wedge shape, increasing in thickness southwards. The sloping upper surface of the wedge has an average gradient of about 1 : 300 although it is steeper near the ground. Rain does not usually occur unless the wedge is at least 5000 ft thick. Rain occasionally falls in the areas covered by the shallow section of the wedge, but more often any cumulus clouds forming here are seen to dry out, and to disappear as their upper parts meet the dry upper air, there being insufficient buoyancy to continue aloft.

In January, the ITCZ is at its furthest southerly position. The associated characteristic rain belt lies to the south of Lagos. South of the rain belt is a zone in which cloud amounts remain high, but there is a considerable reduction in precipitation. This results from a marked inversion, which is at about 3000 ft in the south-east trades, and which is carried into West Africa, where it occurs at about 7000 ft from the south-west, following the deflection after passing to the north of the Equator.

Crowe (1951) remarks that in July, when this 'dry belt' appears in West Africa, the south-east trades are at their strongest, and the waters beneath them are at their coolest, so that the inversion is not dispersed by crossing the Equator, but overrides the strong southerly airstream. This inversion may be penetrated in three ways. The first is by the massive relief feature of Mount Cameroun (13 350 ft). This bodily protrudes through the inversion, and stimulates convective cloud around its flanks. The second is by the more moderate elevations like the Fouta Jallon Plateau, again displacing air aloft, and the third is by surface heating over a continental interior.

The dry belt in the tropical maritime air mass has important consequences. It leads to a 'little dry season' between two rainfall maxima in wet seasons. Also the amount of rain dropped in the southern districts, by the double passage north and south of the wet zone, may be less than the amount dropped where the wet zone arrives, slows, halts and departs. It follows that the maximum mean rainfall totals will therefore tend not to be in the coastal districts experiencing tropical maritime air throughout the year, but in those parts of West Africa which lie a little to the north of the inversion zone at its greatest extent, that is, a little to the north of the transition from a two-peak to a one-peak rainfall regime. (See Fig. 15.8. for the mean monthly isohyets for the month of August). In this zone the average intensity of rainfall may be at a maximum, as not only is the annual total greater than in the coastal regions, but the length of the wet season is less than it is further to the south.

The rainfall experienced at any place thus clearly depends upon its position at the time in relation to the ITCZ, and the seasonal migration of the latter is therefore of the greatest significance. Over the main north–south mass of Africa the zone moves from about to 20° North in July to about 20° South in January, but over West Africa the zone lies north of the coast everywhere east of Cape Palmas. These coastal lands therefore retain their high relative humidity throughout the twelve months of the year, even though in December and January rainfall may be very slight.

At Lagos, the Harmattan breaks through to the coast on only one or two occasions during the year, although 30 km inland it is much more common. Observations lead one to believe there is no ITCZ at this time of the year in this region owing to the shallowness of the humid air, and the general weakness of the circulation. The sea breeze is probably of greater consequence, because its effects can be felt up to 80 km inland. The coastal districts in the early months of the year experience a combination of high temperatures in the order of 32°C with high relative humidity, 85% and more.

North-west of Cape Palmas it is clear that the land in January is north of the ITCZ, and lies under the influence of the tropical continental air mass. Relative humidities along the coast may still be high west of the Fouta Djallon and the Guinea Highlands, of the order 80–85% at dawn, but this is

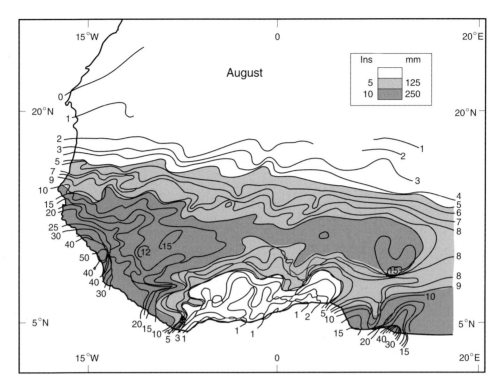

Fig. 15.8 Mean isohyets for the month of August. This diagram clearly indicates the heavier rainfall that occurs north of the coastal regions in this month, with the exception of the eastern coastal area of Nigeria, and the Liberian coast. The main rain belt is now moving south.

principally the result of breezes off the sea, and the protection of the highlands against the dry north-east winds.

In the short distance between Conakry and Boké along the coastal lowlands of Guinea, there is a difference in mean monthly relative humidity (1800 h values) for the first three months of the year: about 70% at Conakry, and 45% at Boké. The lower values continue north-westwards to the southern coast of Senegal, but from Dakar northwards the mean values are slightly higher, due to the effects of sea breezes blowing in from the Atlantic in the afternoons, raising the humidity and lowering the temperature. Strictly speaking, the sea breeze is too weak to overcome the trade winds. The sea breeze merely deflects the trades to blow from a more northerly direction, which makes air movement on the coast slightly on-shore in the afternoons. The morning land breeze effect causes an east to west movement.

The cool Canary current, flowing north to south, is a contributory factor to the early morning mists and fogs along this stretch of coast, most noticeable when the ITCZ is in the area and winds are indeterminate in character. In the areas immediately south of the ITCZ convection rain rarely occurs for the reason stated above, namely lack of depth in the wedge of humid air. The rain that comes is in the form of line squalls, and these will be described.

It is convenient to consider three zones in order to describe the weather in the region of the ITCZ. (See Figs 15.9a, b and c.)

Zone A   North of the ITCZ there is a dry easterly air flow, the Harmattan – a term that is also applied to the dust particles in the air flow. The air is hazy, but visibility below 2–4 km is comparatively rare. Thick haze occurs in spells of a few days at a time, and once or twice each season visibility may fall to 500–1000 m. Visibility is usually best just before dawn, and a marked deterioration sets in an hour or two after sunrise.

Zone B   Just to the south of the ITCZ there is a shallow layer of moist air. The characteristic clouds are small Cu during the day, thin Ac in large amounts about dawn breaking during the morning and variable Ci and Cs. If thick Harmattan haze is present in Zone A, visibility will be poor above the surface in Zone B, and turbulence during the day may lead to a reduction of visibility at the ground to about 8 km or even less. When Zone B lies near the coast (December to February) fog and low St often form at night. Exceptionally, isolated showers or thunderstorms are triggered off over high ground and move east. When the ITCZ is moving north over dry ground these storms may be accompanied by duststorms.

Zone C   A deep layer of moist air enables thunderstorms to develop. Earlier, it was mentioned that the depth of the maritime air must be at least 5000 ft to produce sufficient vertical development of cloud that can give any rain at all. In order to produce thunderstorms, this depth must be at least 10 000 ft. There is much medium cloud before dawn, and low St usually forms shortly after. The latter lifts and breaks during the morning. Later the medium cloud breaks up, by which time Cu begins to develop, to be followed during the afternoon by large Cu and Cb with showers and thunderstorms. These storms are very intense and develop rapidly. They move west at a speed roughly equal to that of the wind (which is from the east) at about 20 000 ft.

This zone is characterised by line squalls which consist of an almost continuous line of thunderstorms lying north–south. They move west at an average speed of about 30 kt. The more active ones are 100–200 nm (185–360 km) in length, and are usually encountered between Kano and Lagos. Most of them occur in the northern part of the zone, and near the coast there are two main squall seasons, one as the ITCZ moves north and the other as it moves south. The tracks of individual squalls can sometimes be plotted over long distances. The West African line squalls are sometimes call tornados. This is incorrect as it is a term inherited from as early as the 16th century sailors, first of all, to describe the fierce squalls in the eastern tropical Atlantic, then applied to the storms seen over land. The name tornado remained in colloquial use until the present day (Kenworthy, 2000).

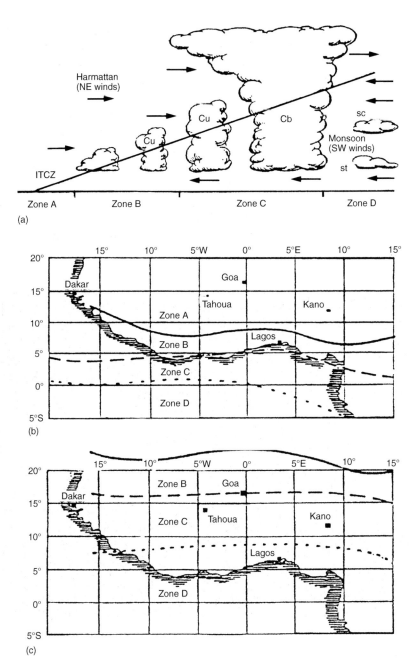

Fig. 15.9 (a) A section through the ITCZ showing the zones associated with cloud development. (b) Mean position of the ITCZ in January (solid line). (c) Mean position of the ITCZ in August (solid line).

A squall is frequently preceded by several days of uncomfortable heat and humidity, and clouds may build up on successive afternoons only to evaporate in the evenings. Once established, the squall moves steadily with intense lightning and thunder, preceded by a few minutes of high winds with speeds up to 55 kt.

The rain front can be seen approaching as a curtain of water, and for the first ten minutes the rainfall is intense, occasionally reaching values of 150–200 mm/h for very short periods. After about ten minutes, the intensity diminishes, and steady rain may fall to give about 25 mm or more of precipitation in the succeeding hour. The rain is accompanied by a fall in temperature. The lateral margins of a squall are also sharply defined, so that while one place receives about 25 mm or more of rain, another place only a few kilometres away may receive nothing. See Fig. 15.10 for a satellite image of a West African line squall.

Fig. 15.10 Satellite image of a West African line squall. (Reproduced with permission from ESA/ EUMETSAT.)

The cause of the line squalls is becoming better understood. The cool equatorial easterlies of higher altitudes play their part. When air is warmed from below and further cooling takes place at higher altitudes due to the overriding cooler air, the lapse rate becomes increasingly unstable. While this in itself can produce intense thunderstorm activity, it is necessary that the wind profiles introduce marked shear in direction as well as speed to make the storm column 'lean' in the direction of its general track. This allows the air ahead of the storm to be cooled by the downdraught. The updraught provides the rain for the downdraught, and the downdraught in turn cools the air ahead of the storm. This cooler air on reaching the ground spills outwards forming a gust front. It is the gust front that causes the further lifting ahead of the storm that provides the feedback mechanism to sustain the storm and for it later to develop into the characteristic line squall. With this mechanism in place, the line squall will be self-propagating, and not affected by diurnal variation. The squall line may continue for over 1000 km. (See Chapter 2, Section 2.8.)

In the dry season similar disturbances may be the cause of 'dust devils', although these are of much less intensity and on a very much smaller scale. The Harmattan and the upper winds both having an easterly component have less interaction, and the extreme dryness of the tropical continental air means that clouds and rain cannot develop unless a very intense disturbance carries the lower air to very great heights. When rain falls from these very high clouds it rarely reaches the ground, but can be seen to disappear on its downward passage into warmer and drier air.

It was mentioned in Chapter 4 that easterly waves may have originated over West Africa. The vigorous line squalls when leaving the Guinea coast could well travel on towards the Caribbean with residual energy to produce a wave. Furthermore, if these disturbances move towards the north-east and cross 5° North latitude, then some of the parameters are in place to spawn an Atlantic hurricane.

Upper winds (Cairo to Johannesburg)

There are the usual two zones of light westerlies either side of the equatorial easterlies.

- In January, the westerly subtropical jet is at 25°–30° North. Speeds in excess of 100 kt.
- In July, there is an easterly jet about 15° North. Speed $\cong$ 80 kt.
- In the south of the route, the light westerlies become moderate, particularly in July.

Tropopause height

North of the route   55 000 ft
South of the route   52 000 ft

Freezing level

From Cairo to Johannesburg 14 000 ft but rising to 16 000 ft across the equatorial zone.

References

Crow, P.R. (1951) Wind and weather in the Equatorial zone. *Transactions & Papers, Institute of British Geographers*, **17**, 33.

Findlater, J.A. (1969) A major low-level air current near the Indian Ocean during the northern summer. *Quarterly Journal of the Royal Meteorological Society* **95**, 362–380. Also personal communication.

Kenworthy, J.M. (2000) The use of the word 'tornado' in West Africa and the eastern tropical Atlantic. *Weather* **55**, February 2000, 60–62.

Chapter 16

Weather in the Middle East

The area covered extends from the eastern Mediterranean to approximately 60°E, covering the Gulf and parts of Iran. The climatic zones are 'temperate transitional' in the north to 'subtropical' in the south. See Fig. 16.1 for the geographical area.

16.1 Flying weather

The area is largely arid. The winter weather is controlled by the Asiatic high pressure, and by the occasional lows which move east from the Mediterranean and pass by way of the Persian Gulf to Afghanistan and northern India.

Away from the coast of the Mediterranean there is little rainfall; what there is occurs in association with the travelling lows and in spring it is often thundery. Over the deserts, widespread duststorms occur with the south or SE winds ahead of the lows.

In summer the area is almost entirely rainless. Over Iraq the NW wind renowned as the *Shamal* is very persistent and carries clouds of dust by day.

In winter the subtropical westerly jet stream covers the Gulf area and much of Arabia; in places the mean speed exceeds 100 kt at heights between about 35 000 and 50 000 ft. In mid summer the low-level NW winds of the Gulf give way to light easterly winds at about 12 000 ft.

16.2 Pressure systems

November to March

The main pressure systems are the winter high over Asia (Fig. 16.2) and depressions that approach from the west. About four depressions each month pass through the Aegean and enter the Black Sea; their cold fronts cross Turkey and secondary lows may form near Cyprus. The depressions that enter Iraq come from the Mediterranean, Egypt or northern Sudan. Some feeble ones originate in Saudi Arabia and have a more northerly track; others form in Iraq under the influence of upper level troughs and jet streams.

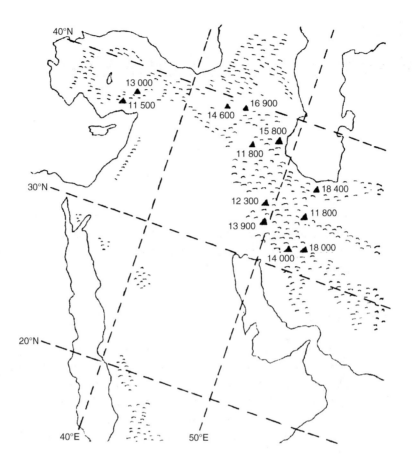

Fig. 16.1 Middle East area topography, heights in feet are approximate.

The frequency of depressions increases in autumn, and from December to March, 6–8 depressions pass east over Iran in each month. While the main ones pass over the western border mountains and cross the country in a somewhat attenuated form, secondaries form over the Gulf. As the season progresses the depressions follow tracks more to the south. They are furthest south in February and March, by which time the frequency of Khamsin depressions in Egypt has increased.

There are a number of variations in the winter mainly due to the passage of depressions following differing tracks (see Figs 16.3 and 16.4). Occasionally, it is to be expected that depressions will cross the area from the Mediterranean. The warm fronts will be weak with little activity but the cold fronts are active especially with orographic intensification over the higher ground. It is also possible for a depression to track further north and arrive via the Black Sea and Caspian Sea bringing the resultant cold frontal weather across the region.

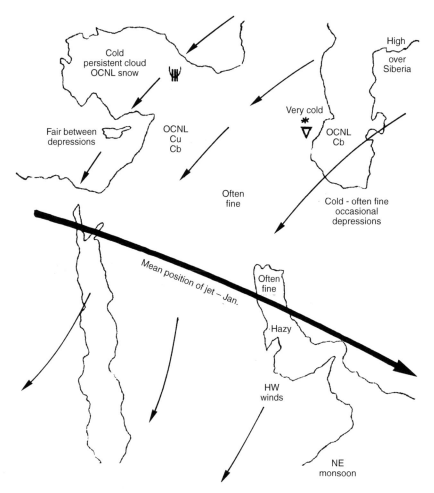

Fig. 16.2 Winter. The Siberian anticyclone predominates with generally north or north-westerly surface winds. Often fine but hazy weather in the Gulf although a risk of early morning coastal fog.

During the winter, the flying weather in the north of the area, over Turkey and eastern Iran is generally rather worse than over the UK. Disturbed weather associated with depressions from the Mediterranean occurs about 50% of the time. Mountain ranges, especially those near the Turkish–Iranian border and in south and SW Turkey, are then enveloped in cloud whose tops may extend well over 20 000 ft. Warm fronts moving from the SE bring widespread precipitation, and on rare occasions a cold front enters Iran from Russia, and bring low ceilings for periods of 3 to 5 days over northern Iran.

In the intervals between the passage of disturbance, layer cloud may persist over the interior of Turkey, but along the Black Sea coast instability, Cu and Sc with showers are very common. In the east, the cloud is more broken, and the highest peaks project above the cloud tops. The Lebanon and Syria may

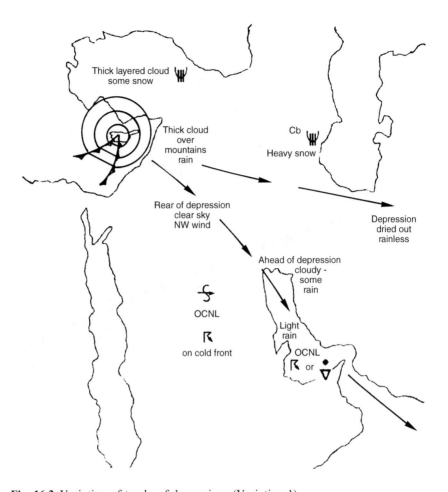

Fig. 16.3 Variation of tracks of depressions (Variation 1).

experience continuous rain when a depression from the west follows a track off southern Greece. Precipitation here and in Israel is generally in the form of heavy showers in this season.

East beyond the mountains the climate gets more arid. In northern Saudi Arabia, Iraq and Iran, conditions are of much the same general character as those of the eastern Mediterranean, but are drier. Warm air occasionally moves across the region from the Gulf, and widespread cloud and rain or drizzle may occur at the warm front.

Cold fronts are marked by line squalls, and orographic effects are responsible for increased cloud on the west side of the mountains in western Iran and also north of Tehran. Thunderstorms occur mainly in this period, but whereas in the west they decrease in frequency during spring, in the east spring is largely the thunderstorm season, when they occur on about three days a month.

There is much clear, fine weather, but early morning fog or low cloud may

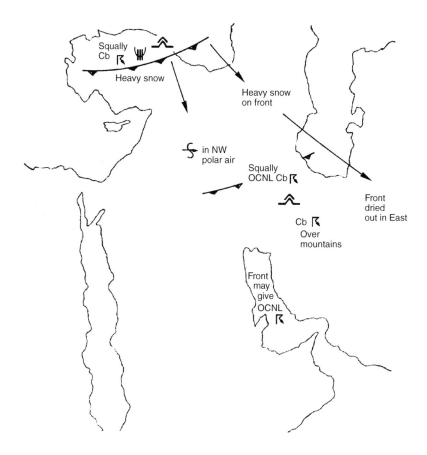

Fig. 16.4 Variation of tracks of depressions (Variation 2). In late winter and early spring, some Mediterranean depressions enter the Black Sea and cross to the Caspian Sea. Cold fronts or troughs form, and trailing fronts can bring Cbs over the mountains.

occur after rain from depressions that cross Iraq, especially in the Tigris/ Euphrates valley. In the interior dust generally rises over a wide area in advance of a depression as the wind freshens, and frequently the dust haze extends up to 10 000 ft.

After the arrival of a cold front, sometimes with thick dust, there is only slow improvement. Over Iraq the frequency of duststorms is greatest between Baghdad and the head of the Gulf; it is higher at Baghdad than at Basra, and March is the worst month. Most but not all duststorms occur during the day; some even persist overnight. On the day following a severe duststorm, the visibility is often reduced below 1000 m although the wind is insufficient to raise dust. In the Gulf and east along the coast, NW squalls (*Shamals*) sometimes accompanied by duststorms and thunderstorms are associated with the cold fronts of western disturbances. Starting in March dust is blown about by moderately strong winds during the day. In the region of the Gulf this is most marked on the west side where the visibility is often poor. There may be some fog or low St around dawn in the Gulf area.

Summer – June to September

In general, clear skies are common and rainfall generally small to the north and west of the region and negligible to the east. Over Turkey and Iran there may be some patches of low cloud in the morning and small amounts of fair weather Cu in the afternoons. However, along the Black Sea coast the formation of Cu can be considerable; heavy showers and occasional thunder occur, and Istanbul has most of its thunderstorms from May to August. Cu/Cb also form over the highlands of eastern Turkey and NW Iran.

Occasional duststorms occur over the Turkish plateau and reduce visibility below 1 km. Near Tehran there is often dust haze up to a height of 10 000 ft (see Fig. 16.5).

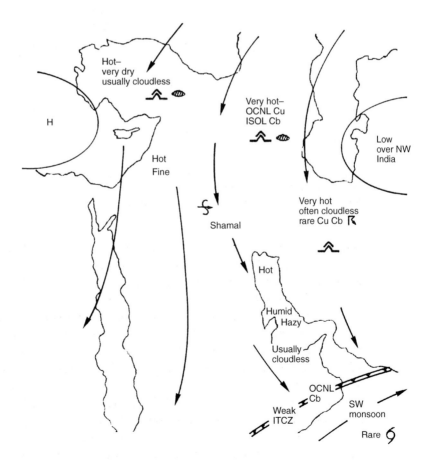

Fig. 16.5 Summer wind-flow circulation.

The wind-flow circulation is now controlled by the airflow around the low over southern Iran and north-west India. It is mainly hot, humid and usually cloudless. Visibility can be reduced in the northerly wind (*Shamal*) if dust is lifted, which will usually be on a diurnal basis. The ITCZ or even a tropical cyclone may affect the southern end of Arabia but this is not usual.

16.3 Upper winds

Winter
Mainly westerly with the subtropical jet stream axis aligned east–west across the region. Typically 260°–280°T/110 to 140 kt.

Summer
The temperature gradient has lessened and the subtropical jet stream is not evident. Winds generally light easterly. Typically 090°T/15–30 kt.

Height of freezing level

Winter   8000 to 12 000 ft
Summer   16 000 ft

Tropopause height

34 000 to 56 000 ft

Visibility

It is usually hazy over the Gulf, but sand or dust can reduce visibility at any time with winds in excess of 16 kt. Sandstorms occur frequently in summer with the north wind (*Shamal*).

Chapter 17

Weather – Arabian Gulf to Singapore

This area is largely dominated by the monsoons of Asia. In winter, when the Siberian high is established, the interior of the continent becomes a source region of polar continental air. The dry and cold air flows out over the surrounding areas. In summer, the high pressure gives way, and a low to the NW of India draws in moist equatorial air from the Indian Ocean. Thus, the direction of the low-level circulation in summer is largely the reverse of that in winter, and spring and autumn are transitional seasons in which there is no definite circulation.

The climatic zones are a combination of 'subtropical' and 'equatorial'. The word monsoon means 'season' but it is applied in other parts of the world that experience a very wet season.

As the monsoons are most important for the Indian subcontinent they are listed as follows:

Winter   –   NE monsoon (cool season) (Dec/Feb)
Spring   –   Inter monsoon (hot season) (Mar/Jun)
Summer   –   SW monsoon (general rains) (Jun/Oct)
Autumn   –   Inter monsoon (cyclone season) (Oct/Dec)

There is a variation to some aspects of the weather as Singapore is approached and this will be described later.

During the NE monsoon (December to February) the mountains to the north shelter the area from the direct outflow of the continental air. Weather is predominantly fine with little cloud and good visibility, but the shallow depressions which occasionally move east from the Mediterranean bring some cloud and rain. Also, the coast near Madras tends to suffer from low cloud off the sea.

The winds flowing down the Ganges valley are turned to become the NE monsoon of the Bay of Bengal. Owing to the long passage over the warm sea, moisture is picked up with the result that the coast of Madras and the NE of Sri Lanka now have considerable rain with cumulus and cumulonimbus. This produces thunderstorms with a maximum activity about dawn. Fog sometimes occurs in the early morning over low-lying areas but it clears soon after sunrise. Occasional thunderstorms may occur in Bangladesh towards the end of the winter monsoon period.

Occasionally, depressions, which have originated over the Mediterranean or Iran, may penetrate into NW India. These lead to a temporary breakdown of the monsoon conditions. The numbers and paths of these depressions vary considerably from year to year, probably depending on the position and intensity of the Siberian anticyclones. In the NW of the area the winter rain is associated only with the passage of these disturbances (see Fig. 17.1).

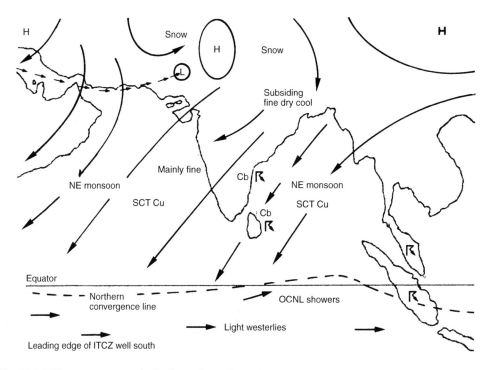

Fig. 17.1 NE monsoon over the Indian subcontinent (December–February) → → → → Track of rare 'Mediterranean' depressions – give the only rain in north-west India (snow on mountains).

In the transition period (or inter-monsoon period) (March to June) the weather is mainly hot, dry and dusty. The weak lows which continue to traverse northern India from west to east are at this time associated with severe squalls and duststorms.

These storms tend to become more frequent as the season advances. A feature of this season is the thermal lows which develop over the Thar Desert leading to very high temperatures and quite frequent sandstorms and duststorms over Sind and Balochistan. Severe wind squalls are associated with these as with the thunderstorms (see Fig. 17.2).

Meanwhile, the pressure begins to fall in NW India and the SE trade winds cross the Equator to feed the SW monsoon, which lasts from about June to

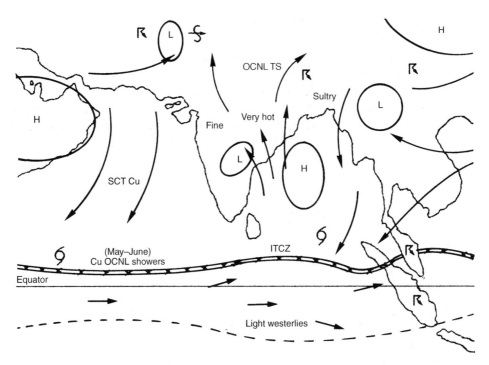

Fig. 17.2 Hot weather season over Indian subcontinent (March to June).

October. The onset of the monsoon takes place suddenly at any one place. It reaches the west coast of India in early June and then extends NE. After its long sea passage the air is moist and convectively unstable, and heavy orographic and convective rain become widespread. The monsoon cloud and rain just reach the coast of Balochistan, but further west and in the Gulf, conditions remain arid.

Tropical cyclones form in the Arabian Sea and Bay of Bengal during the advance and retreat of the monsoon, that is, when the ITCZ lies over these seas. Other cyclonic storms form over the Bay during the monsoon. Some of these move NW and give copious rain in the Ganges Valley. Here the air stream is SE, but as it progresses inland it dries out and NW India gets little of the monsoon rains.

Flying conditions in the SW monsoon are difficult on account of the severe turbulence, the massed clouds and rain and the obscuring of high ground. However, visibility is good outside the rain and cloud, the bad weather tends to be patchy and some days are better than others, even to the extent of 'breaks' in the monsoon. After the retreat of the monsoon during September, the weather generally improves. (See Fig. 17.3.)

During the second half of September the SW monsoon is retreating south. Winds are generally light and variable with thunderstorms at times but less

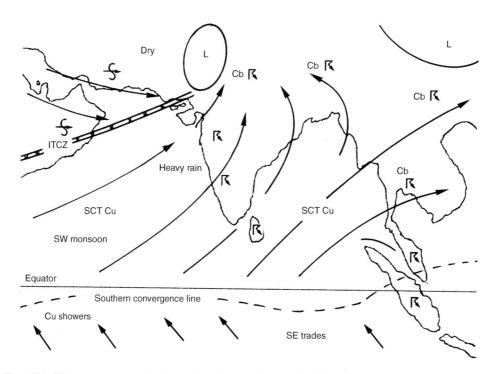

Fig. 17.3 SW monsoon over Indian subcontinent (June to October).

severe than those of the SW monsoon period. In the north fine weather is soon established and this fine weather spreads gradually south until by December it covers the whole of the subcontinent. This season is, however, the season of maximum cyclone activity, the Bay of Bengal being most affected. Most of these storms move north towards the Ganges valley, but associated with them is a wide area of cloud and rain which affects coastal areas of Madras. Tropical cyclones also occur in the Arabian Sea but they are rather less frequent. These storms also occur in the season March to June (see Figs 6.4 & 17.4).

17.1 Tropical depressions and cyclones

Low-pressure systems occur in the Bay of Bengal and to a lesser extent in the Arabian Sea. The term 'depressions' is used for those in which the wind does not reach 35 kt. More vigorous systems are called cyclones. On average about 12 depressions occur in the Bay of Bengal each year, of which 6 develop into cyclones. The depressions, which form just before and after the SW monsoon, especially after, are much more likely to develop into cyclones than those that form during the monsoon season, April–May.

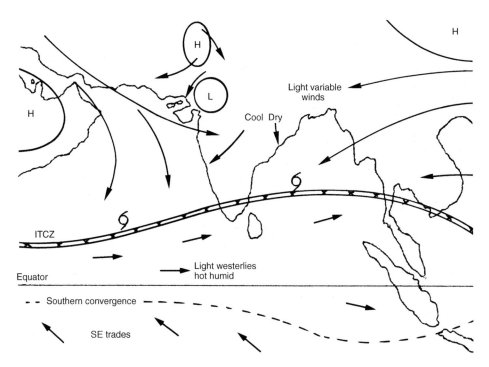

Fig. 17.4 Autumnal climate of Indian subcontinent (October–December).

The coastal regions affected and typical tracks vary with the seasons.

Jan–Mar	SE India and Sri Lanka, from the east and SE.
April–May	Mainly Myanmar (Burma) (north of Yangon (Rangoon)) and Pakistan from the south and SW.
June–Sept	NE India and Bangladesh, from the south and SE in June, and from the east in July to September when some have travelled from the China Sea. Tracks are much more irregular by September.
Oct–Dec	No part of the coast is immune, but Bangladesh and the Indian coast near Madras are the areas most affected. Some come from the China Sea and cross lower Thailand. Tracks are very irregular, but usually they are westerly, later curving to the north and NE.

In the Arabian Sea the frequency is much less, two to three each year on average. The most likely periods are May–June and October–November, but even then the frequency is less than one a year in each period. The coastal region most affected is between Bombay and Karachi, but the majority fill up over the sea.

In the last week of May or the first week of June the SW monsoon is usually ushered in by a depression from the Bay of Bengal crossing NE India. During the monsoon season a trough extends across the Ganges valley, and depressions (about five) cross northern India. During July and August they reach Pakistan. As the monsoon withdraws depressions still cross the country, but their paths get progressively further south; in December they are confined to the extreme south. From October to December some of them may enter the Arabian Sea.

17.2 Climate of Malaysia

This area is subject to the SW and NE monsoons, but the weather associated with both is very similar. One of the outstanding features of the climate is its monotony. Generally the sky is almost cloudless at dawn, but cumulus begins to form around 0900 hours and increases in size and density as the day wears on. The clouds grow to cumulonimbus during the afternoon, this development resulting in rain and thunderstorms. Towards dusk the cloud decreases and the sky is generally clear by midnight. Cloudless nights are, however, much less frequent on those coasts which are exposed to the current monsoon. All through the year the same daily cycle takes place, with scarcely any variation to distinguish one day from the next.

It is true that the rain-gauge and the sunshine records indicate that at the change of the monsoons there is generally a little more rain and a little less sunshine than at other times of the year, but to a casual observer the insignificant seasonal differences are barely noticeable. No well-marked travelling depressions invade the area, and only thunderstorms and squalls of comparatively short duration disturb the otherwise uneventful daily cycle of the weather.

The worst weather and cloud conditions occur from about mid-September to December when the SW monsoon is being replaced by the NE monsoon. Conditions then are almost always poor with overcast skies and a great deal of rain along the NE coast. Much of this weather spreads to the western side during the afternoon, but the mornings are usually fair. Cloud is very persistent over the hills, especially on the eastern slopes, and the hilltops are frequently obscured.

January to mid-March
The NE monsoon is in complete control. Weather along the NE coast is showery, but the northern half of the west coast has much fine weather, owing to the sheltering effect of the mountains inland. Most of the rain in the south occurs in the afternoon or early evening.

Mid-March to May
This is a transition period of mainly light and variable winds over the whole

area. Showers may occur at any time, but most mornings are fair and any bad weather usually occurs later in the day.

Mid-May to September
This is the season of the SW monsoon. Over most of the country the weather is better than at any other season because:

(1)   The monsoon air current has usually originated over Australia and is usually somewhat drier. Please note: the winds are SE south of the equator and swing to become SW to the north.

(2)   Sumatra, with its extremely extensive mountain ranges provides a sheltering effect. This effect is not noticeable along the northern part of the west coast and showery conditions prevail there.

October to December
This is another transition period and the SW monsoon is retreating southwards. Conditions are generally similar to the previous transition period. Tropical storms are extremely rare, owing to the proximity of Malaysia to the Equator.

Sumatras

These are squalls which occur during the season of the SW monsoon. They originate as a line of extensive cumulonimbus clouds over the mountainous backbone of Sumatra during the afternoon and advance across the Malacca Strait during the night as a line of thunderstorms. A katabatic wind forms on the slopes of the mountains and the storms move out over the Straits of Malacca. It is accompanied by a squall which may attain a speed of 40–50 kt. Sumatras may be expected on average once every 10 days, but tend to occur on 2 or 3 days in succession when they do commence.

Sumatras may occur simultaneously along a line as much as 350 km in length, running roughly NW–SE, aligned with the Straits of Malacca, the whole squall travelling in a northerly to NE direction with a speed corresponding to that of the squall wind. Sumatras may temporarily intensify when they cross the west coast of the Malaysian Peninsula, the ground now warming up with the onset of the sea breeze; however, they are seen then to lose their vigour and die out. They are seen more frequently in the north than in the south of Malaysia and so the frequency at Singapore is relatively low. See Fig. 17.5 showing a satellite image of 'Sumatra' cloud formations.

17.3 Summary

In January with the ITCZ over the Equator the NE monsoon covers the whole area. In July with the ITCZ to the north of the region the SW monsoon covers

Fig. 17.5 Satellite image of Malyasia showing 'Sumatra' clouds in the Straits of Malacca. Image taken in the early morning before the clouds moved onto the Malaysian mainland. Photo: NOAA VIS: HRPT, downloaded by Chris van Lint in Hong Kong and supplied by Peter Wakelin (RIG). Reproduced with permission. Acknowledgement to NOAA.)

the whole area. In the spring inter-monsoon period winds are generally northerly in the west, light and variable over India and NE over Malaysia. In the autumn winds are NW over the west of the region, light and variable over India and light westerly south of the ITCZ northern edge.

Upper winds (300 mb)

January	270°/60–90 kt
July	090°/15–30 kt
In Singapore	light easterlies

The axis of the subtropical jet stream moves rapidly from south of the Himalayas to the north in early June coinciding with the start of the SW monsoon season.

Tropopause height

January	57 000 ft
July	54 000–55 000 ft
Singapore	56 000 ft (all the year)

Visibility

Generally there are few visibility problems except for duststorms in the desert areas in the west of the region. Fog and mist often occur in river valleys, early in the morning in the tropical areas.

Cloud/precipitation

Almost all cloud will be cumuliform with typical heavy shower type precipitation. Monsoon conditions may be experienced with cloud tops anywhere between 25 000 and 50 000 ft.

Freezing level

Usually between 15 000 and 18 000 ft. Severe icing is probable in any towering cumulus and Cb.

Chapter 18
Weather – Singapore to Japan

The area under consideration is from the Equator (Singapore is only just north of the Equator) to 40°N, and 100°E to about 145°E. The climatic zones are equatorial in the south, subtropical high through temperate low pressure in the North. See Fig. 18.1.

This area comes under the influence of the Asiatic monsoons with two brief transition periods. The ITCZ lies across the area in summer. Its position is ill defined, but in July it often lies east of Korea to southern Japan. As the winter monsoon sets in so the ITCZ retreats, eventually to well south of the Equator.

18.1 November to March

This is the winter period dominated by the NE monsoon which when fully developed forms a stream of cold continental air flowing from the NW across northern China towards Japan and thence curving back to cross southern China and Vietnam/Thailand from the ENE (see Fig. 18.2). There are thunderstorms and heavy rain in the Singapore area (ITCZ well south). A series of cold fronts form in the north of the route, as cold air crosses the sea and the warm Kuro Shio current. Convective cloud develops and it can be orographically enhanced over coastal areas and mountains of Japan. Precipitation includes snow on the western coasts, but on the more sheltered east coast of Japan it can be much warmer with a risk of lee depressions developing.

In the more sheltered area on the SE side of Japan, such as the Tokyo Plain, the skies are often clear. The most common cloud is small amounts of afternoon Cu, but on occasions dense cloud penetrates gaps in the mountain range, especially round Nagoya, and shower clouds sometimes reach the east coast in the north. Over the sea the weather is cloudy (mainly Sc) with occasional showers or light rain.

A front separates the monsoon from the NE trades. It may be anywhere between southern China and northern Malaysia. When inactive the front has a broad belt of drizzle or slight rain, with low St and poor visibility to the north of it, but these conditions only affect the Ryukyu Islands when the front is

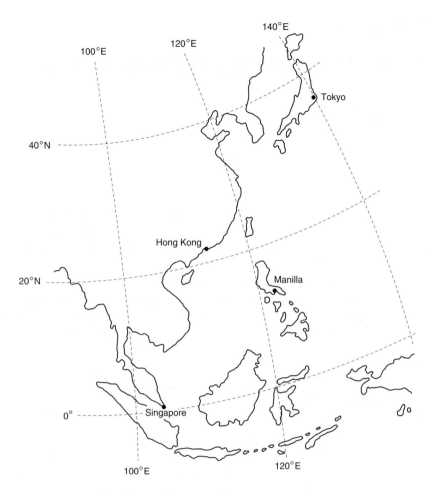

Fig. 18.1 Geographical area.

north of its typical position. The worst weather in this area occurs when waves form on the front. Many of them develop west of Taiwan and move towards the east or NE.

About three days in five have a full monsoon. A lull begins with the formation of a trough of low pressure over northern China or further east between the Siberian high and a detached anticyclone, which moves east into the Pacific. Waves form on a new polar front in the trough and as each wave moves east the cold front sweeps further south.

Over mainland China and around Hong Kong it is cold and dry from October to December, but from January to April the *Crachin* can apply. This is a relatively local feature of SE China caused by interaction between Pc air and Tm air from the sea. It gives low St, drizzle and fog along the coastal area and it can penetrate inland. Frontal precipitation is widespread, with considerable snow or sleet in the north of Japan and drizzle or rain to the south;

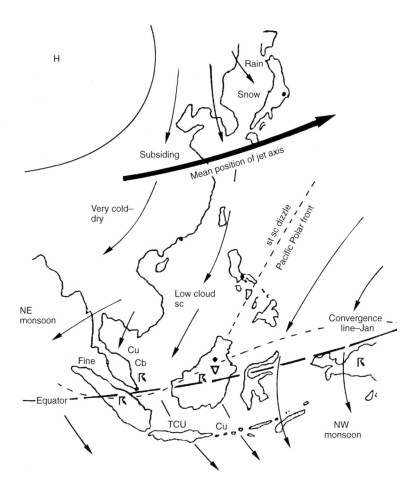

Fig. 18.2 Air-flow patterns (winter) November to March.

warm sectors are cloudy and misty. There may even be precipitation over the whole of Japan when an intense low lies just off the SE coast and a lee depression forms over the East Sea of Japan.

The winter monsoon is about 5000 ft deep and westerly winds prevail above 10 000 ft. On average their strength increases with height up to about 35 000–40 000 ft. Their constancy is high (frequency greater than 70% near Tokyo), and very strong winds are at times also observed, $\cong$ 250 kt (even 300 kt has been observed) in the upper troposphere and more than 150 kt at around 18 000 ft. *The strongest mean winds in the world are found in this region.* The subtropical jet stream is normally near 30°N and almost always from a direction between 240° and 290°T. Further north lies the polar jet stream, *which from time to time moves south to reinforce the subtropical jet stream.*

18.2 April to August

The Siberian high breaks down in late March and is replaced by a succession of anticyclones and depressions which move east into the Pacific (see Fig. 18.3). The cold fronts of some of these depressions penetrate into lower latitudes, often reaching southern China. Further south the SW monsoon is spreading north. Depressions form on its leading edge, especially when a cold front crosses southern China, and moves east. Their main path is close to the Ryukyu Islands in May and early June, which in this area is the wettest and cloudiest period of the year. Rainfall in the depressions may be heavy and

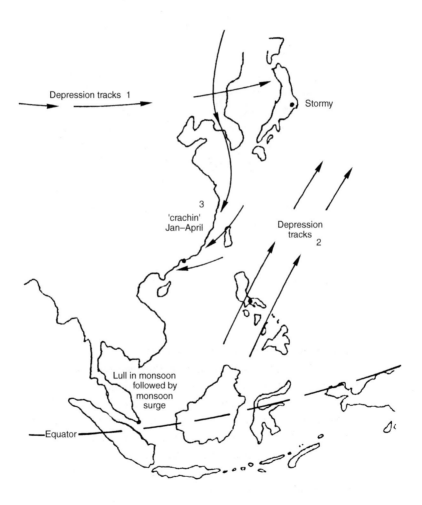

Fig. 18.3 Lull in the monsoon. Depressions can form on the Pacific front and move NE (Track 2). Note that depressions can cross Asia from Europe (Track 1). Track 3 produce the 'Cratchin' on the China mainland.

widespread, and over Japan the SE slopes are wetter than the NW. Thunder becomes common towards the end of the season.

The SW monsoon continues to move north as the North Pacific subtropical anticyclone intensifies. By late July this warm, damp air stream extends over much of Japan. Over the sea the weather is mainly fair, with scattered Cu; over the land large Cu and Cb develop during the day with severe afternoon and evening thunderstorms.

During June and early July the depressions along the leading edge of the SW monsoon move across Japan. The sky is then overcast with frequent heavy rain (called the Bai-u or Plum rains). When they reach southern Japan, easterly winds over the cold coastal waters (the Oya Shio current), bring sea fog to the east coast, mainly north of Tokyo (see Fig. 18.4). At the same time warm air rises over the easterly winds and heavy rain may fall for hours.

A trough of low pressure frequently lies in a SE direction across or just north of the Philippines. It is the main breeding ground for typhoons, which occur with increasing frequency during this season. The ITCZ lies a little to the east of the trough. On occasions, the trough is absent and the SW winds of the China Sea extend into the Pacific. This tends to occur when a typhoon is moving NE towards Japan.

The disappearance of the subtropical jet stream described in the last section coincides with the establishment of the SW monsoon over India. This is followed a little later by the disappearance of the polar jet stream, and by July the upper westerly winds are confined to an area around Japan.

18.3 September to mid-October

Pressure is now rising over Asia, and cold fronts begin to penetrate south across northern China and Japan, bringing behind them several days of cool, dry northerly winds, the first burst of the NE monsoon. Such a burst is most likely to occur when a typhoon or tropical depression is moving away NE. By October it usually reaches as far as Vietnam and the Philippines.

Vigorous depressions form on the cold fronts over northern China and move east over northern Japan. Each frontal passage brings heavy and prolonged rain to Japan. The SW monsoon and the Philippines trough are still frequent in early September and also occur in October. Typhoons continue to form in the trough, but their frequency of occurrence decreases as the season progresses. By October the high-level westerly winds are established over the area above 18 000 ft and easterly winds may occur near Hong Kong. The polar jet stream has reappeared, and winds of more than 100 kt are likely from time to time near Japan.

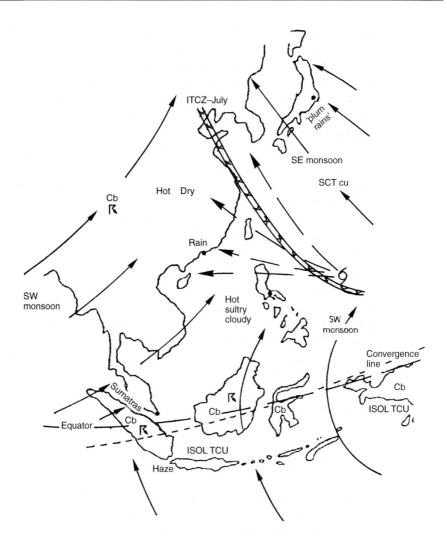

Fig. 18.4 Summer wind-flow patterns (April to August).

18.4 Typhoons

About 70% of typhoons form in the Pacific well to the east of the Philippines; some also form over the South China Sea. Most of them occur in late summer and early autumn.

The main Pacific group originates near the ITCZ. Many of these move slowly WNW towards Taiwan or the Ryukyu Islands, and then recurve NE towards Japan, but some continue to move NW towards the China coast. The South China Sea group is erratic in its movements, but a common track is NW towards Vietnam.

Typhoons usually move at speeds between 5 and 15 kt, and once located on the weather chart their movement can generally be predicted with sufficient accuracy to enable adequate precautions to be taken.

18.5 Summary

Tokyo weather

In winter warm fronts associated with depressions off the south coast of Japan give snow and a low cloud base below 1000 ft. Visibility can be reduced below about 800 m by persistent smoke particularly during the morning and evening when the winds are light.

In summer, typhoons occur once or twice a month from June to November. Fog and smoke haze is frequent from April to October in the morning when there are light southerly winds.

Upper winds

In winter, the axis of the subtropical jet stream extends from China to southern Japan. The monsoon is overrun at about 7000 ft to give westerly winds. Mean speed over Japan in January is about 150 kt at 40 000 ft (200 kt + not unusual). Moving south the winds become light westerlies and then light easterlies in the Singapore region. In summer the subtropical jet stream disappears and upper winds are light westerly to the north of the route and light easterly in the south.

Ocean currents near Japan

There are two significant ocean currents near Japan, and reference is made to them in various parts of the text. See also Fig. 2.3. The *Kuro Shio* current is a warm current. It is a fast current of about 2–4 kt, flowing *northwards* from Taiwan to the Ryukyu Islands and close to the coast of Japan to about 150°E. Also known as the Japan current.

The *Oya Shio* current is a cold current flowing from the Bering Sea *south-wards* along the coast of Kamchatka, past the Kuril Islands, and continuing close to the north-east coast of Japan. It is detectable to nearly 35°N.

Visibility

Fog can occur in early mornings in the south of the region. Japan is prone to advection fog in winter with morning fog and smoke haze common in April to October in association with light winds.

Freezing level

January   Singapore 16 000 ft   Japan 3000 ft
July   Singapore 17000 ft   Japan 10 000 ft

Tropopause height

Singapore   All year   56 000 ft
Japan   Winter   35 000 ft
            Summer   42 000 ft

Chapter 19
Weather – Singapore to Australia

This area lies in the Equatorial trough and the subtropical high pressure climatic zones, although influence of the subtropical high pressure is only effective at the margins, mainly May to September.

The ITCZ moves across this area from about 15°S in January to 20°N in July. Also, there are tropical cyclones in the south, January to March. The passage of the ITCZ is a major contribution to the weather, but there are other convergent zones that appear due to the wind being deflected by the Indonesian islands. This results in local convergent zones and can appear particularly when the ITCZ lies in the south of the area, but they can form in any season. The route Singapore to Darwin is shown in Fig. 19.1.

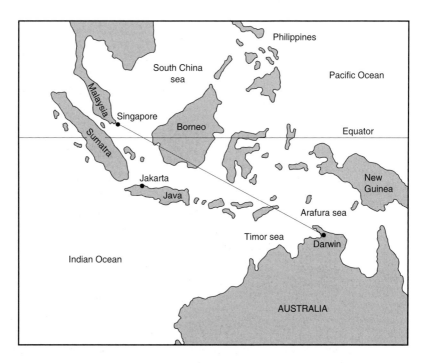

Fig. 19.1 The route, Singapore to Darwin.

This area has an equatorial rainy climate except in the northern part of Australia. In January the NE monsoon reaches the area. The moist and unstable air, after a long sea track south of the Equator, veers to become the NW monsoon of northern Australia, which then lies within the Equatorial trough.

In July, Australia is in the belt of subtropical high pressure. SE trades blow from the continent and from the Pacific Ocean and veer to become the SW monsoon north of the Equator. Generally there is a great deal of convective cloud with frequent heavy showers and thunderstorms. However, the topography of the islands is extremely varied and there are often marked contrasts of climate between the windward and lee sides.

Land and sea breezes are a regular feature. That part of the SE trade wind, which originates over Australia in the southern winter, is dry and dusty over the continent. It is associated with a period of haze and reduced rainfall in the islands south of about 5°S. Generally the wettest periods at any place occur when the ITCZ is in the vicinity, that is, in one or other of the transition seasons.

Tropical cyclones develop in or near the Timor Sea from January to March. They usually move at first SW and after recurving pass into NW Australia; they are accompanied by heavy rain and strong winds or gales.

19.1 Movement of the ITCZ

The ITCZ is the boundary between the air streams originating in the northern and southern hemispheres. If these winds do not converge to any great extent, the zone is merely a wide area of ascending and descending air currents often referred to as the 'doldrums', within which areas of showers and squalls are interspersed with regions of bright, clear weather and calms. Where convergence between the trades is extreme, the zone may appear as a well-defined squall line with cumulonimbus clouds building up to above 40 000 ft or more.

The zone is not always continuous. It may be strong in some areas and at the same time weak and diffuse in others. It may appear as one or as several squall lines, and when present as a wide 'doldrum' area may have two distinct 'edges,' both active, with the doldrum area of light winds in between. On some occasions the whole of the area between the edges may consist of bad weather, but such widespread deterioration is rare.

During the advance of the monsoon, the zone is often pushed rapidly forward by surges of the monsoon air. As the flow weakens, the zone will dissipate, and reform to the rear again in advance of a new surge of air. The day-to-day movement of the inter-tropic zone is very erratic. It often remains almost stationary, but has been known to move 150 to 200 km in a day.

The boundary between the monsoons is not always clearly defined. In particular, over narrow landmasses and islands it may become a diffuse, broad zone in which convection storms over the land are more frequent and persistent. The zone may also be broad, but with feeble convection when the air streams are weak. More vigorous development then takes place along its north and south edges, to form what appears as two boundaries. The zone is most likely to be well defined during rapid movement north or south.

The ITCZ is present over the route during the months October to June inclusive, and is virtually clear of the region during the months July to September, when it is near its northernmost position. Figure 19.2 shows the position of the ITCZ in January, when the leading edge of the zone is in the neighbourhood of Darwin. Figure 19.3 shows its position in July, when it is virtually clear of the route.

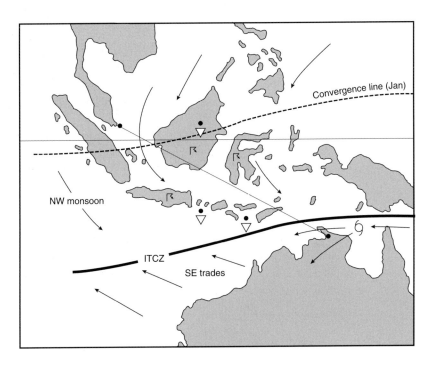

Fig. 19.2 ITCZ location in January. The NW monsoon dominates the route, until near Darwin, where the SE trade winds meet at the ITCZ.

The ITCZ arrives over Singapore and Malaysia during October, ahead of the strengthening NE monsoon; it reaches its southernmost position through Darwin in January and February. It is interesting to note that the direction of island chains and the distribution of land and sea affect the alignment of the ITCZ.

The zone commences its northward advance in March, and by June has moved north of Malaysia to a position through Thailand and the central

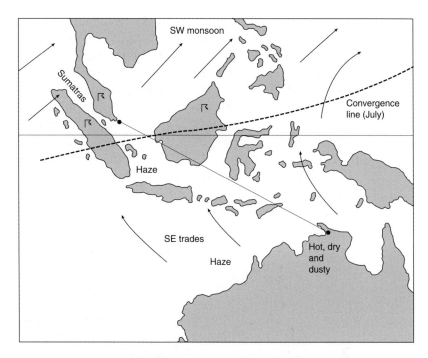

Fig. 19.3 The ITCZ in July has advanced to the north of the chart. The route is virtually clear of the zone, and the SE trades dominate. The wind directions will veer after crossing the Equator to become the SW monsoon in the northern hemisphere.

Philippines. The movement of the ITCZ is affected by the activity of pressure systems outside the tropics. Surges of the northerly monsoon or the passage of pressure systems in the southern hemisphere cause the position of the zone to fluctuate. The presence of a typhoon further north or off NW Australia usually means it moves rapidly.

19.2 Other convergence zones

Lines of convective activity may also be encountered outside the ITCZ. Some develop in winter (southern hemisphere) from cold fronts, which have advanced into the area. They are aligned east–west. Many are due to topography. These can form in any season and are aligned in any direction.

Topographical convergence zones (sometimes called orographic fronts) are a mesoscale weather phenomenon. They result in convergence within an air mass and their formation is dependent on the direction of the air stream and the distribution of the land and sea, or the alignment of mountain ranges. They are common on this route because of the numerous islands. Two examples are shown in Figs 19.4 and 19.5. Once convergence zones are initiated, cumulus,

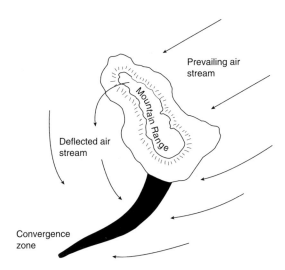

Fig. 19.4 Convergence zone forming from orographic deflection (example 1).

towering cumulus and even thunderstorms result. There would then be no diurnal variation if the winds continue to meet in the convergence area.

When an air mass crosses the Equator, part of the stream is sometimes undiverted and part will change into a westerly stream. This may happen as the result of topography, see Fig. 19.6. The undiverted air steam is channelled southwards off the east coast of Borneo, and meets the westerly air stream off the southern coast. This then forms a convergence zone.

Convergence zones may form without the assistance of adjacent land-masses. The 'local' convergence zone so produced often forms a *triple point* with the ITCZ (see Fig. 19.7). Such a mechanism is a source of extremely bad weather; it would only form in the hemisphere which is experiencing its summer. In Fig. 19.2, the position of the ITCZ is shown north of Darwin. The

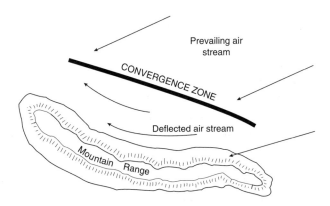

Fig. 19.5 Convergence zone forming from orographic deflection (example 2).

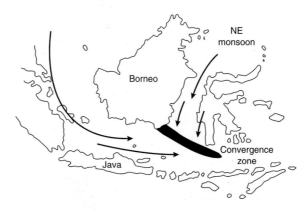

Fig. 19.6 Convergence zone formed from undiverted and diverted airflows.

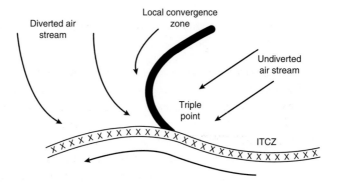

Fig. 19.7 Local convergence zone formed without orographical assistance. Undiverted and diverted air streams meeting and forming a triple point.

ITCZ can move to the south (see Chapter 21 for further details about the weather in Australia).

19.3 Thunderstorm patterns and land/sea breezes

In many coastal areas, especially where the direction of the monsoon is from sea to land, a relatively strong rainfall is observed in the latter part of the night. This is due to showers and thunderstorms forming over the sea at night and being carried across the coast by the onshore breeze during early morning. At Jakarta, for instance, morning showers come from the north, but the afternoon showers come from the south. Morning thunderstorms are also experienced at Darwin when north-westerly winds are present.

The thunderstorms triggered by convection show a well-marked daily variation with a maximum in the afternoon. Thunderstorms are formed earlier at the mountain tops and on the slopes than in the lowlands, which is in accordance with the fact that these showers may originate on the slopes and afterwards descend to the plains. Records show that rainfall maximum (during the inter-monsoon periods when the majority of the precipitation comes from local thunderstorms) occurs at 1400 or 1600 hours LMT in the hill country and at 1700 to 1800 hours at coastal stations.

The phenomenon described above is a very regular one over all sections of the area. Day after day, clouds are formed during the morning hours in the mountains and rains occur in the afternoon. The reverse takes place over sea areas. A general feature over the whole of the region (assuming no local convergence zone activity is present) is the absence of extensive low cloud formations over the sea during the daytime. The sea water remains relatively cool during the hours of daylight, and so the effects of insolation and vertical displacement are absent. Conditions are smooth for flight at lower levels, and generally the low fair-weather cumulus encountered is present in scattered patches only.

However, at night, when a stable stratification of air is being established over the land, turbulent vertical currents arise over the sea. The loss of heat by long wave radiation from the water vapour laden upper air layers causes them to become cool relative to the warm sea surface, with resulting instability. It is common at coastal stations to observe, at sunrise, a line of towering cumulus and cumulonimbus cloud to seaward often with lightning. Turbulent cumulus over the sea disperses rapidly after sunrise, and calm smooth conditions are soon restored.

Land and sea breezes develop in coastal regions with monotonous regularity, and when the monsoons are weak these diurnal wind circulations predominate. Even when the monsoons are well developed, their direction near coasts changes considerably in the course of the day, due to the influence of the land and sea breeze components. The change from land to sea breeze occurs at about 1000 to 1100 hours LMT, and then from a sea breeze to a land breeze shortly after sunset near mountainous coasts, but it is later when the coastal areas are flat.

19.4 Flying weather

November/December to March

The northerly monsoon lasts from November to March in the east, and from December to March in the west. See Fig. 19.2 for the position of the ITCZ and wind flow. The predominant clouds are Cu and Cb, and there is also a great deal of medium cloud. Well inland the Cu and Cb reach maximum development in the late afternoon, and during the night they flatten and degenerate to layer clouds which may persist near mountain tops while fog and low St may form in valleys. Over the sea the time of maximum development is the night and early morning. As a result coasts often have a secondary maximum of rainfall during the second half of the night. The number of days with heavily clouded skies is very frequent, and windward slopes are very often covered in cloud. Regarding fog and/or mist in river valleys, this is a characteristic virtually *throughout the year* in these regions.

Rainfall usually occurs in torrential downpours lasting for short periods, but along disturbance lines, and in particular along mountain ranges at right angles to the monsoon wind flow, rain may be prolonged. The visibility in heavy rain may fall to almost zero. Thunderstorms and rain squalls are frequent over land, but there are wide variations from place to place.

Favoured localities have as many as 25 days of thunder each month, but sheltered areas no more than 10. Thunderstorms form mostly on inland slopes and drift towards the coasts in the afternoon and evening. Over the sea they are less frequent, especially in the western half of the area where they are isolated, and are carried towards the land by the monsoon during the night and early morning.

Cyclones occur between Timor and Darwin, most of them in the period January to March, but they are infrequent. The southerly monsoon lasts from May to September in the west and from April to October in the east.

April/May to September/October

Near Singapore the sky is only slightly less cloudy than during the northerly monsoon, but cloud amounts decrease towards the SE. From Java SE there is little cloud, mainly small Cu over land in the afternoon, but there are some large Cu and Cb at the beginning and end of the season. On occasions a band of cloud associated with a weak convergence zone lies east to west over the Timor Sea.

Haze reduces visibility to 15–20 km up to about 8000 ft in the east and 6000 ft near Jakarta. Further west, the haze is less dense and not so common, but in the east it occasionally reduces visibility to 5–8 km.

The frequency of thunder decreases to the east. Near Singapore there are more than 20 days in May, 10–15 in other months. In the region of Jakarta the numbers are roughly half as great. Beyond Java thunder is rare in the middle months and occurs on a few days early and late in the season.

Rain squalls are quite common near Singapore. *This is the season for the 'Sumatras'.* A weak type of cyclone occurs in April and early May. It moves towards the SW past Timor, but is very infrequent.

Inter-monsoon seasons

These periods vary in time of occurrence and duration. In the west they occur about April and October–November. In the east they usually last less than a month and occur about March/April and October/November. Cb clouds are well developed, and in the west there are numerous rain squalls and thunderstorms, the latter particularly on inland slopes but also over the sea if the ITCZ is active.

Upper winds
Most of the route will experience light easterly winds at all times of the year, typically 090°/15–30 kt. In the Darwin area there will be light westerlies in January and westerlies up to 60 kt in July. No jet streams.

Tropopause height
On average, ranges between 52 000 and 56 000 ft over the whole route.

Freezing level
Usually around 16 000 ft over the whole route. Severe icing may be experienced above this level in large Cu and Cb.

Visibility
The south-east trades blowing over Australia bring dusty, hazy weather with generally poor visibility. In tropical areas fog and mist form overnight in river valleys, clearing soon after sunrise.

Chapter 20

Weather in the South-west Pacific Region

The following section includes extracts taken from the New Zealand Meteorological Service (Misc. Pub. No. 166, 1980, The climate of the south-west Pacific region, compiled by J T Steiner) and reproduced with the kind permission of the Ministry of Transport, New Zealand Meteorological Service.

20.1 The summer weather systems

In the eastern part of the area, the inter-tropical convergence zone (ITCZ) is always north of the Equator. A second convergence zone, the South Pacific convergence zone (SPCZ) lies in a north-west to south-east orientation in the southern hemisphere (see Fig. 20.1).

The position and intensity of these features vary considerably. The SPCZ occurs where the south-east trades on the northern sides of the eastward-moving anticyclones shown in Fig. 20.1 meet the air flows which originate in the semi-permanent anticyclones of the South-east Pacific. This current is divergent over the island groups just south of the Equator where there is a remarkable zone of low rainfall. Part of the air flow moves as a north-easterly onto the northern side of the SPCZ.

There is a shallow depression, *a heat low*, over north-western Australia. A convergence zone is shown extending eastwards across northern Australia to the southern Solomon Islands. (The convergence zone does not always pass through the heat low centre as shown in Fig. 20.1). This convergence zone forms part of the ITCZ.

The continental ITCZ is rather different from oceanic convergence zones. The air to the south of the zone is usually rather dry and often cloudless, whereas the moist monsoon current to the north of the convergence zone is generally cloudy. Thus, over Australia the ITCZ appears not as a cloud band as over the ocean but as the southern limit of the monsoon cloud mass. To the south of the convergence zone, the trade winds blow from the east over Queensland. The onshore trades, the monsoon and the occasional cyclone bring *summer rainfall* to northern Australia. The north-west monsoon airflow

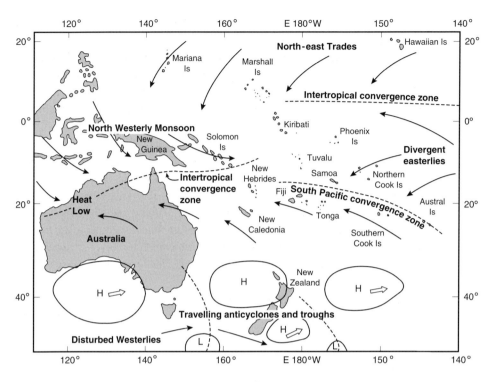

Fig. 20.1 Typical circulation in the South-west Pacific area – January (summer).

usually extends eastwards to the Solomon Islands, although occasionally it may extend as far east as Tuvalu.

In the pattern illustrated in Fig. 20.1 the two parts of the ITCZ are not linked. On some occasions there may be a continuous ITCZ across the South Pacific. More commonly the western ITCZ is continuous with the SPCZ. Another common occurrence is to find no well-organised link but rather an area of generally light winds (doldrums area), which displays weak convergence, and convective cloud in the region between the convergence zones. Frequently, the doldrums area is just east of the Solomon Islands.

A series of anticyclones with their centres at about the latitude of New Zealand is shown moving generally eastwards. It is common for an anticyclone centre to slow down west of New Zealand while a ridge of high pressure extends eastwards. A new anticyclone centre subsequently forms east of New Zealand and then moves east while the old centre loses intensity. (This is the case in Fig. 20.1.)

Anticyclones, as they move eastwards, have a tendency to also move equatorwards. This is not an invariable rule and the northward component of the motion does not, of course, continue indefinitely. The strength and size of an anticyclonic circulation changes from day to day, growing to a maximum and then decreasing. Finally the anticyclone disappears. Often, as an anticyclone collapses, a new one forms somewhat to the south-west. The cycle of growth, movement, decline and disappearance is thereafter repeated.

There is a transition from winds from the northerly quarter to winds from the southerly quarter as the trough between successive anticyclones passes. North of the latitude of the anticyclone centres the wind change in summer is often quite gradual and may be unaccompanied by any significant changes in cloud or by any precipitation.

To the south of the anticyclone centres the transition from relatively warm north-west winds to colder south-westerlies, with the passage of a trough, is not generally gradual or smooth. It is often quite abrupt, and accompanied by rain and sometimes squalls and thunderstorms. This narrow transition zone is a cold front. Rarely, a further sharp transition occurs, usually only in the more southern latitudes, when colder air in the rear of one trough is replaced by warmer air in advance of the next one. These transition zones are warm fronts. More frequently warming between successive cold fronts takes place slowly by subsidence.

The troughs of low pressure are associated with cyclones (or depressions) centred in high latitudes. The depressions also go through a life cycle of growth and decay as they move generally east or SE. As depressions decay in high latitudes they are replaced by new cyclones originating from lower latitudes.

Tropical and mid-latitude cloud systems are often connected. The SPCZ cloud sheet is commonly continuous with middle latitude frontal bands. Similarly there is often a continuous band of cloud extending from the northern areas of Australia (the heat low and ITCZ area) to frontal belts over the southern half of the continent.

20.2 The winter weather systems

In Fig. 20.2 it can be seen that the ITCZ lies in the northern hemisphere at all longitudes. It is generally furthest north in the west of the region. The southern hemisphere trade winds cross the Equator and extend well into the northern hemisphere. The SPCZ is often present and has approximately the same orientation as in summer. Occasionally in winter the SPCZ is absent and a broad undisturbed east to SE flow prevails over the island groups.

It will be noted that the south-easterly flow is shown to extend over the Solomon Islands and New Guinea. There is, however, usually a decrease of wind speed in this area with the consequent convergence and upward motion resulting in convective activity. This area of convective activity occurs at about 10° South and sometimes extends east to about 160° West.

The usual latitude of the anticyclone centres is a little further north in winter than in summer in the area east of New Zealand. In the Australian region the anticyclonic centres are appreciably further north in winter. The variation in typical latitudes of anticyclone centres in this region is partly caused by the

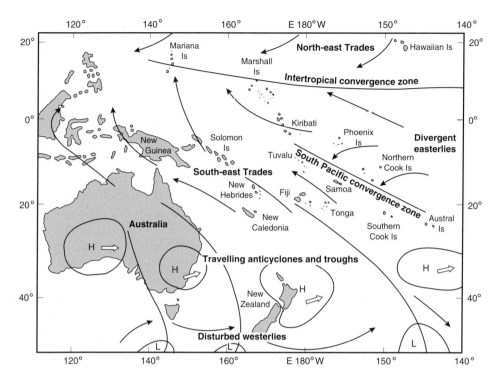

Fig. 20.2 Typical circulation in the South-west Pacific area – July (winter).

seasonal variation of the land–sea temperature difference. In the vicinity of New Zealand the seasonal variation in latitudes of anticyclone centres is quite small. In winter, as in summer, some distortion in the shape of anticyclones usually occurs as they pass over New Zealand. An example of this is shown in Fig. 20.2.

The cycle of anticyclone development, eastwards with some northwards movement, and decay is otherwise similar to that of summer. The sharp frontal transitions generally extend further north in winter than in summer. Many southern coastal regions of the Australian mainland, for example, receive significant rainfall from winter troughs, whereas summer troughs frequently pass over with no precipitation.

When the anticyclone centres are further north than usual over the Tasman Sea/New Zealand area, disturbed westerlies often extend over much of New Zealand with rapid changes from troughs with belts of rain to ridges with fair weather. Occasionally, the usual pattern of anticyclones in the north and depressions to the south is disrupted. Anticyclones moving towards the east or NE from the area south of Tasmania are preceded by an outbreak of cold southerlies. Such higher latitude anticyclones often move very slowly and the cold weather on their eastern side can therefore be rather persistent. Another important variation in the winter pattern is the development of deep slow-moving lows over the north Tasman Sea.

20.3 The spring and autumn weather systems

The typical spring and autumn patterns may be generally described as intermediate between those already considered, with spring more closely resembling winter, and autumn more closely resembling summer. However, it is noteworthy that in spring the anticyclone centres in the Tasman Sea/New Zealand region are usually appreciably further north than in other seasons. Consequently disturbed westerlies commonly occur over New Zealand in spring.

Upper winds

Upper winds are considered for the 500 mb ($\cong$ FL 180) and 200 mb ($\cong$ FL 380) levels. The 500 mb level is in the middle troposphere, and the 200 mb level is within the upper troposphere in tropical regions but may be in either the troposphere or stratosphere in the latitude of New Zealand.

The main anticyclonic belt at 500 mb in January is at about 10° to 15° South, but in the west there is an anticyclonic circulation over central Australia. The high temperatures in the lower troposphere over the Australian continent in summer give rise to large, 1000–500 mb thickness values and hence to the anticyclonic circulation at 500 mb. In the region north of the anticyclones, easterly winds are found over the equatorial region.

At the 500 mb surface over eastern Australia, the upper air contour lines are shown to be cyclonically curved, corresponding to an upper trough. Elsewhere, in the middle latitudes, the flow is generally westerly. The strongest winds are between 45° South and 50° South (south of Australia).

The 200 mb level chart for January over the region shows anticyclonic centres over the Northern Territory and at about 10° South to the north-west of Fiji. These centres are further north than at 500 mb. In the vicinity of the Equator, the winds at the 200 mb level are easterly in the west of the region, and westerly in the east, the changes in direction occurring near the date line. In the middle latitudes the flow is generally westerly. There is a very slight cyclonic curvature in the contour lines over eastern Australia. The strongest winds, exceeding 50 kt, occur between 45° and 50° South.

Westerlies increase rapidly south of the anticyclonic belt, and in the central and eastern parts of the region, and reach a maximum of more than 50 kt at about 25° South, decreasing to light winds in the latitude of New Zealand. In the longitude of Australia there is rather less variation in the wind speed between latitude 20° and 60° South.

The anticyclonic centres at 200 mb in July are slightly further north than in

January. East of the centre, NE of Fiji, there is a region in which the contour lines are anticyclonically curved extending from the Tokelau Islands to just north of the Society Islands. The equatorial wind flow is rather more complex than in January but again there are westerlies in the east and easterlies in the west. The strongest westerly winds south of the anticyclonic belt occur at 25° to 30° South with winds in excess of 100 kt over Australia and to the north of the Tasman Sea. At most longitudes, winds are relatively light at about 45° South but increase again further south.

A point to note is that the areas of anticyclonic circulation at the 200 mb level in both January and July correspond quite well to the zones where cloudy conditions are most frequent. These are also the regions where low-level convergence is common. Thus the cloud of the SPCZ is associated both with low-level convergence and with upper-level divergence leading to anticyclonic circulation aloft.

The strong wind belts at high levels are the subtropical and polar jet streams. The 200 mb level is identified because the strongest winds in the subtropical jet stream are always close to this level. The strongest winds in the polar jet stream occur at different levels in different seasons. In summer and autumn the mean level is near 250 mb ($\cong$ FL 340), but in winter and spring the winds at high latitudes continue to increase with height to well above 100 mb ($\cong$ FL 540). This high-level, high-latitude mean wind maximum is known as the *polar night jet' stream*.

The description of upper winds so far has shown that the mean flow at upper levels is almost zonal. When maps for individual days are examined the flow over the whole region is never as zonal as described. On some occasions the flow is mainly zonal but with small amplitude ridges and troughs so that the meridional wind component is much less than the zonal one. On other occasions large amplitude troughs and ridges occur with strong meridional flow. Furthermore, the jet stream location on individual days may be quite different from the mean locations described.

In winter and spring, a continuous jet stream extends across the South-west Pacific from Western Australia to beyond the longitude of Tahiti. Maximum jet stream winds flowing at more than 150 kt are quite common. A spot wind of 250 kt has been measured on the Auckland to Nandi route.

Jet streams (southern hemisphere observations) – note
(See also Chapter 3, Sections 3.1 and 3.2)
Vertical and horizontal gradients of wind speed, i.e. shears, are strong in the vicinity of the jet stream. The areas of strong shear are often turbulent. The turbulence is occasionally severe. Severe turbulence probably occurs in only small regions and may well be a rather transitory phenomenon. For these reasons it is difficult to forecast.

There are favoured areas for upward and downward motions associated with the jet stream. Upward motion tends to occur upwind of the jet maximum on the equatorial side of the jet and downwind of the jet maximum on the polar side. If there is sufficient moisture in these areas cloud will form. The amount of upward motion is increased in areas where the jet is blowing from the north-west, i.e. between a trough and the next ridge downstream. Once the air passes the ridge axis, downward motion tends to occur leading to cloud dissipation. Thus it is common to find extensive cloud sheets associated with the jet stream originating in the region of north-westerlies and dissipating hundreds or even thousands of kilometres downwind where the flow turns south-westerly.

The strong vertical shear of the wind immediately below the jet stream implies the presence of a strong transverse horizontal temperature gradient. Thus the region below the jet stream can be considered as an upper level front. The cloud sheets associated with the jet stream are, however, different in their mode of formation and their structure from those associated with surface frontal zones. They are caused by ascent over a wide area rather than by ascent in narrow zones of convergence.

Tropical cyclones note
In Chapter 6, tropical storms have been described. In the South-west Pacific area, these storms are called cyclones. A point that needs clarification is that the Beaufort scale describes winds of force 12 as 'hurricane force' (winds exceeding 64 kt). Winds of hurricane force do not only occur with hurricanes. They also occur with storms of different characteristics, such as tornadoes, which are appreciably smaller than hurricanes and can produce extraordinary wind speeds. Also, some mid-latitude storms can have a greater area of strong winds than do hurricanes and they, unlike hurricanes, do not have a warm core.

Tropical storms (southern hemisphere observations) – note
(See also Chapter 6, 'Tropical Storms')
Small cyclonic disturbances are very numerous particularly in the southern summer between latitudes 5° South and 20° South and to the west of 160° West. Of the numerous circulations in this area, only about four or five per year develop into severe storms, although in some years as many as twelve potentially destructive storms have occurred.

Cyclonic disturbances are uncommon on the Equator side of 5° South. Those that occur south of 20° South and east of 160° West, or develop in the winter, rarely develop into severe storms as they move away from the tropics into higher latitudes.

When the surface winds in a tropical storm of the *South-West Pacific* are force 12 on the Beaufort scale, i.e. at least 64 kt, the storm is referred to as a hurricane on the Beaufort scale. (Tropical storms of this intensity in the *North-west Pacific* are called typhoons.) Because *hurricanes are most likely from November to April*, this part of the year is referred to in some of the Pacific island groups as the 'hurricane season'. However hurricanes have on occasions occurred in other months of the year. Hurricanes are likely to contain areas of severe turbulence and should be avoided by aircraft at all levels. Thunderstorms and torrential rain are additional hazards.

It is commonly said that tropical cyclones generally move initially towards the west or south-west (north-west in the northern hemisphere) and that they 'recurve' between latitudes 20° and 30° to follow a path towards the SE (NE in the northern hemisphere). Authors qualify this generalisation by recognising that much more complex tracks are common and that not all tropical cyclones recurve in the manner described. In the South-west Pacific area, however, the exceptions are more common than the rule. A study of the tracks of tropical cyclones in the area in the years 1940 to 1969 showed that 10% proceeded west without recurving, 30% did recurve as per the 'rule', 35% proceeded east without recurving, and 25% proceeded east and recurved towards the west (Kerr, 1976). Thus, in the South-west Pacific in the period studied, only two-fifths of the cyclones moved westwards initially, and almost half of the total number of storms did not recurve at all.

When the storms move into temperate latitudes south of 30° South and many do, they generally move SE, but this, too, is not an invariable rule. Some move southwards or even SW and, of course, changes in direction and speed of movement occur in these latitudes also. The storm also changes from a warm core storm to a cold core storm. (See Chapter 6, 'Tropical Storms'.)

The ITCZ and SPCZ (southern hemisphere observations) – note
(See also Chapter 5, 'The Inter-Tropical Convergence Zone (ITCZ)')
The ITCZ varies in width from about 50 to 500 km. It varies considerably in intensity both from day to day at any fixed longitude and from place to place on any day. In general when it is wide, the convergence zone is also active, when narrow it is weak. When it is weak the ITCZ comprises scattered cumulus clouds associated with scattered layer cloud. There is also isolated cumulonimbus, possibly with anvil tops.

It is difficult to describe the cloud of an active ITCZ as viewed from the ground in terms of the conventional cloud types. This is because both cumuliform and stratiform clouds are meshed. There is apparently a dense stratiform sheet, though the presence of areas of heavier rain with low ragged clouds suggests that there is also convective cloud. Viewed from an aircraft at $\cong$ FL 380, numerous bubbly cumuliform tops are usually seen emerging out of a dense cloud mass. There is also extensive layer cloud. A few very large cumulonimbus clouds may extend upwards to very high levels, probably to $\cong$ the tropopause. The huge cumulonimbus clouds are accompanied by severe thunderstorms.

High stratiform and cirroform cloud is seen extending outwards from the ITCZ in the direction of the flow in the middle and upper troposphere. Satellite (visible) imagery of active convergence zones shows large bright white clusters. They are typically of somewhat irregular shape and a few degrees of latitude across. The bright almost uniform surface whiteness is due to the cirrus cloud cover. On some occasions cumuliform turrets may be evident. The ITCZ moves very little from day to day.

The SPCZ has a similar cloud distribution to the ITCZ; like the ITCZ, it is sometimes active and sometimes weak. In winter the SPCZ is often weak or it may not even exist. The movement of the zone is irregular. Often it may appear in one area for a few days, then weaken, to reappear in another area; on some occasions two distinct convergence zones appear. The convergence zone cloud sheet often appears to be reactivated with the passage of a cold front in middle latitudes. In fact the SPCZ and the middle-latitude cold front often appear as an almost continuous cloud band.

The SPCZ has associated with it a trough in the sea-level pressure pattern. The pressure often falls in some areas of the SPCZ so that a closed surface isobaric depression develops. When this occurs, the SPCZ visually expands in width and becomes very active.

On other occasions shallow depressions form in the tropical South-west Pacific at a distance from any existing convergence zone. With the formation of the depression a new convergence zone forms through the region of the depression. The resulting weather pattern is generally quite similar to that of depressions that form on existing convergence zones.

The giant cumulonimbus clouds of the ITCZ play a very important part in the atmospheric circulation. The updrafts of these clouds provide a means of taking energy to high levels. This process 'drives' the tropical circulation.

References

Kerr, I.S. (1976) Tropical storms and hurricanes in the South West Pacific. *New Zealand Meteorological Service Miscellaneous Publication* 148.

Steiner, J.T. (1980) The climate of the South West Pacific. *New Zealand Meteorological Service Miscellaneous Publication* 166.

Chapter 21
Weather in Australia

The weather for northern Australia is influenced partly by the position of the ITCZ. The main landmass is under the influence of the subtropical high pressure belt. There are travelling depressions across the southern part of the continent. See Fig. 21.1 for a map of the Australian subcontinent, and the principal air mass tracks into and out of the continent. The general weather over Australia can be seen in Fig. 21.2.

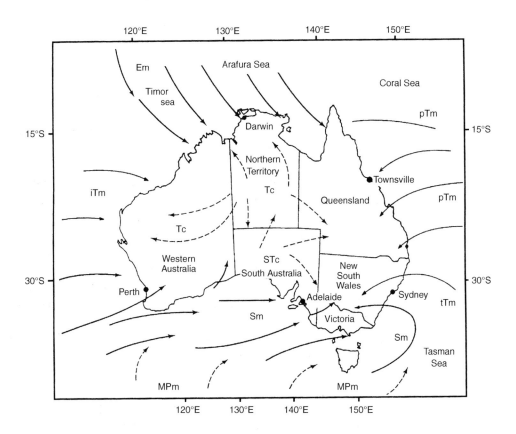

Fig. 21.1 Map of the Australian subcontinent, and the principal air mass tracks.

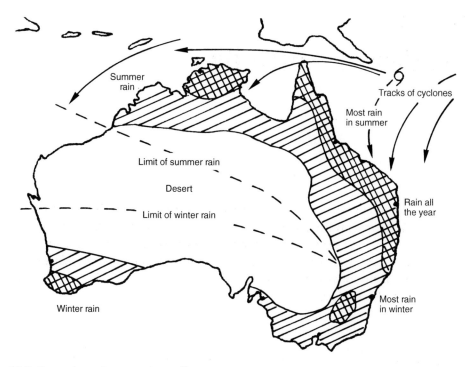

Fig. 21.2 General weather over Australia.

21.1 Air masses affecting Australia

There are eight major air mass types that affect various parts of Australia. These are as follows.

(1) *Modified polar maritime (MPm)*   This is a very cold, moist and unstable air mass. It originates in the Southern Ocean on the margin of the Antarctic (55–68°S) and is characterised by dew-point temperatures of 2 to 7°C. It only affects southern Australia occasionally in winter (during strong southerly flow after the passage of a vigorous cold front) and is often accompanied by snow and sleet to low levels.

(2) *Southern maritime (Sm)*   This air mass originates in the Southern Ocean (35 to 55°S), and is a cool air mass and moist (dew-point temperatures 7 to 13°C). It is unstable at low levels, but is stable above. It brings cool, moist, cloudy weather and drizzle to southern Australia at any time of the year, but little rain unless orographically displaced.

(3) *Tropical maritime Tasman (tTm)*   This is a warm air mass and originates in the north Tasman Sea. It is unstable and moist to high levels. The dew-point is high (13 to 18°C), and the air mass brings warm weather with cloud and drizzle to coastal regions of eastern Australia, with heavier rain if orographically displaced. This air mass is present along the

central coastal regions most of the year, but its effects diminish towards the south, particularly in the winter.

(4) *Tropical maritime Pacific (pTm)*   This air mass is similar to the tTm air mass but is warmer. It originates further north in the Coral Sea and tropical western Pacific Ocean. With a dew-point of 18 to 21°C, this air mass affects the northern Queensland coast most of the year. Rainfall can be heavy if associated with a tropical cyclone.

(5) *Tropical maritime Indian (iTm)*   This air mass has very similar characteristics to pTm. It originates in the Eastern Indian Ocean and affects the north-western coastal areas of Australia.

(6) *Equatorial maritime (Em)*   This air mass originates over the ocean areas to the north and west of Australia. It is very warm and moist (dew-point temperatures 21 to 24°C) and unstable. It only affects north and north-western Australia in summer in association with the monsoon. It brings extremely heavy rainfall and high humidity to this area, but during very active monsoon seasons can affect areas as far south as 30°S.

(7) *Tropical continental (Tc)*   This air mass originates over central Australia. It is very hot, dry and unstable in summer, but is cooler in winter. Dew-point temperatures are low, typically from −4°C to +2°C. Cloud and rainfall are severely limited by a lack of moisture. A further limiting influence is the trade wind inversion in the mid-troposphere, which effectively inhibits upward instability. This air mass affects the north-central areas of Australia for most of the year. On occasions, with strong northerly winds, heat wave conditions will be experienced in southern Australia.

(8) *Subtropical continental (STc)*   This air mass originates over south-central Australia. It is warm and dry, and dew-point temperature is 2 to 7°C. It is the dominant air mass over inland southern Australia in winter. This air mass forms in the high pressure belt (descending air in the Hadley cell circulation).

21.2 North Australia

The ITCZ lies close to the north of Australia in the southern summer (see Fig. 21.3), but in the southern winter lies far to the north. Hence, the south-east trades and the north-westerlies mainly control the weather, and there are two distinct seasons, a wet season (November to March) and a dry season (April to October).

Diffuse zones of convergence are more usual than sharply defined fronts. Such zones may form in any area where air streams have a component towards each other. During the dry season the cold fronts, which cross the continent between anticyclones, are of little importance, except perhaps for dust. During the wet season they seldom advance much further north than 20°S and active

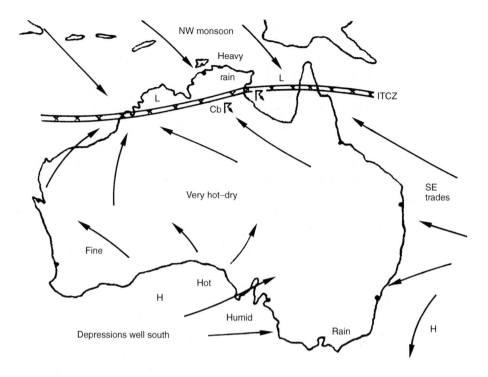

Fig. 21.3 Position of ITCZ in January. North-west monsoon in the Darwin area (southern summer).

severe conditions exist through the entire width of the zone, which may be 100–130 km wide. The ITCZ is present only in January and February. Normally it is north of Darwin but occasionally it moves south. When active, it may take the form of a broad convergence zone, with the central portion often fairly clear, although heavy cloud and weather may be experienced through the entire zone. See Fig. 21.3 for the position of the ITCZ in the southern summer.

A further variation of the ITCZ in northern Australia can be seen in Fig. 21.4. The 'triple point' is formed where the NW monsoon to the north is meeting converging air masses drawn in by the boundaries of the ITCZ and a frontal zone crossing the mainland. There is high pressure towards the south and SE. The 'local' convergence zone extends towards the south. This results in extremely bad weather with Cb and thunderstorms.

21.3 South Australia

In this area, frontal depressions are frequent in all seasons, but the Eastern Highlands greatly modify the weather so it is very different inland to that on the coast. Cold fronts move east across the continent in troughs between

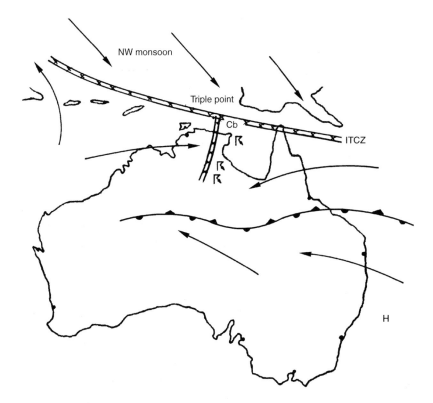

Fig. 21.4 Position of the ITCZ in January showing a 'local convergence' zone.

anticyclones. With moist air from the Pacific ahead of each front there are extensive clouds, heavy rain and thunderstorms. With dry air the front passes with very little cloud, but in summer it is often marked by violent duststorms in the interior.

Waves may form on these fronts, but mainly over the sea to the south and SE. When they do form over the land they bring extensive low cloud and rain over SE Australia and southern Queensland. Such a wave may persist for several days, and others may form on the original front and move south-east.

The cold front often trails back to form a warm front in advance of the next cold front, mainly in the southern summer. The trailing front becomes more active when it lies inland across Queensland and the warn air which is moist produces a great deal of cloud which may cover the entire state.

Tropical cyclones

The area most commonly affected by cyclones lies north of 20° South, and the season lasts from December to April. The majority of cyclones occur in January, February and March. The greatest number originate north of

Darwin in the Arafura or Timor Sea, with a frequency of about five or six each cyclone season.

Not all cyclones are fully developed, but even a small centre may produce extensive cloud and bad weather. They normally move SW along the coast, but some cross the coast of the Northern Territory. Others come from the Coral Sea. They rarely enter the interior, but those reaching the Gulf of Carpentaria often rejuvenate before passing between Darwin and Daly Waters, to bring widespread cloud and rain. Some curve south over the Pacific, but rarely affect the coast as far south as Sydney.

November to March
Near Darwin, the skies are often heavily clouded. Cyclones and shallow lows bring widespread cloud, which may persist for several days. In general, the cloud development is greatest during the day. In bad conditions the cloud base may fall almost to the surface, especially in rain squalls; Cb is almost a daily occurrence. From about 20°S to the Eastern Highlands there is only a little more cloud than in winter. Daytime Cu and Cb can be very marked, especially in Queensland, and severe afternoon thunderstorms are sometimes experienced. However, bad weather areas are generally isolated.

Conditions vary beyond the Eastern Highlands. Away from fronts, bad weather is normally confined to certain air streams from the sea. The main ones are from the SE after a long track over the Tasman Sea, and from the south when they are moving rapidly and are unstable.

Thunderstorms are most frequent near Darwin (10–15 days of thunder each month) where they may be extensive. Those that occur in the evening may remain active until after midnight, even as far north as 20° South, and some of those over the sea may drift over the coast near Darwin just after sunrise.

In the interior, thunderstorms are infrequent and amount to no more than three each month near 25° South. Along the East Coast the frequency decreases towards the south. Brisbane has six or seven each month until February, Sydney slightly fewer, and Melbourne no more than two.

Line squalls occur mainly in the southern half of Australia, and inland they are often accompanied by severe duststorms. Most of them arrive from the south with the advent of a cold front. Squalls frequently follow in the airstream behind the front. A particularly vigorous line squall affects the coastal strip of New South Wales (*Southerly Buster*), but it is unknown west of the Highlands. Five to six a month are likely from November to February, two-thirds of the annual total. Surface winds can be very gusty after the passage of the squall line; 70 kt gusts have been recorded. Temperature changes can be dramatic, particularly if the squall line passes through the region during the afternoon. A fall in temperature of 10°–15°C in a few minutes is quite common. The leading edge is associated with a spectacular roll cloud; when this occurs, it is orientated perpendicular to the coast. However, clouds are not always present, and the abruptness presents a hazard to low flying aircraft.

Turbulence near the Highlands can be severe. See Fig. 21.5 for the wave that causes the 'Southerly Buster' (spring and summer). Note that the wind from the interior is called the '*Brickfielder*' – which is hot and dry.

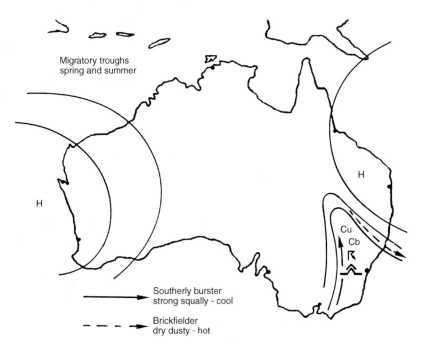

Migratory troughs
spring and summer

H

H

H

Cu
Cb

Southerly burster
strong squally - cool

Brickfielder
dry dusty - hot

Fig. 21.5 The wave formation causing a line squall ('Southerly Buster'). The brickfielder is a hot dry wind from the interior.

April to October
From Darwin as far as the Eastern Highlands the skies are often clear or cloud is scattered (Fig. 21.6). Near the northern coast there are some large Cu and Cb early and late in the season, and just inland, near Daly Waters, low St occurs in the early morning. The SW air stream may bring cloud and showers to the western slopes of the highlands.

In the SE, the sky is a little less cloudy than in summer because of the greater frequency of westerly winds. Thunderstorms are infrequent and occur mainly near the beginning and end of the season; in the interior they are rare. The possibility of mountain waves near the Eastern Highlands must be borne in mind, as with the McDonnell Ranges near Alice Springs.

Duststorms occur inland, and in Queensland and the Northern Territory they are more frequent in this season. From May or June large quantities of smoke and ash are carried NE, and thick haze from dust and smoke is generally present. The haze extends up to a height of 6000–8000 ft. Visibility may fall below 3 km and the sky may be obscured. Cyclones may affect Darwin in April (see Fig. 21.2 for cyclone tracks).

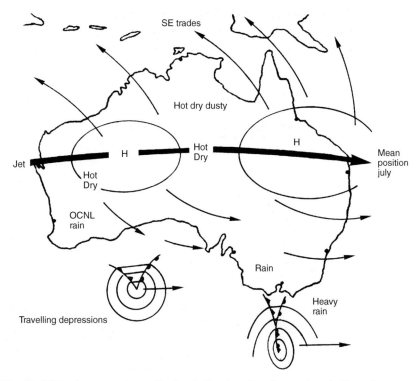

Fig. 21.6 Weather over Australia April–October (southern winter). Note the mean position of the subtropical jet stream.

21.4 Summary

November to March

Surface wind at Darwin	North-westerly
Upper wind at Darwin	Generally westerly
Subtropical jet axis	Across the south of the continent
	Westerly 60–70 kt at 200 mb
Surface temperature	20–30°C all the year
ITCZ southern limit	Approximately 15°S
Surface wind at Sydney	Variable
Upper wind at Sydney	Generally westerly
Weather	Moist SE winds provide extensive low cloud cover
Surface temperature	19–29°C

April to October

Surface wind at Darwin	Easterly – light
Upper wind at Darwin	Generally westerly

Subtropical jet axis	Over the centre of the continent – 100 kt Westerly at 200 mb
Surface temperature	20–32°C all the year
ITCZ	North of Singapore
Surface wind at Sydney	Variable
Upper wind at Sydney	Generally westerly
Weather	Early morning fog possible. Smoke haze possible below the anticyclonic inversion. Probable clearance with a sea breeze
Surface temperature	9–15°C

Freezing level

	Winter	Summer
20°S	10 000 ft	16 000 ft
40°S	6000 ft	12000 ft

Tropopause height

	Winter (July)	Summer (Jan)
20°S	53 000 ft	55 000 ft
40°S	34 000–49 000 ft	38 000–51 000 ft

Heights refer to both Australia and New Zealand.

Chapter 22
Weather in New Zealand

Some of the following notes have been taken from CA Pamphlet No. 28 *'Meteorology Study Notes for Commercial Pilots'* published by the Civil Aviation Division, Ministry of Transport New Zealand, and incorporated with kind permission.

22.1 The climatic situation

New Zealand is influenced by the subtropical high-pressure belt in the southern summer, and travelling depressions in the southern winter. The climate of New Zealand is influenced essentially by three main situations, which are as follows.

(1) The oceanic environment, the topographical features of the land and the fact that it lies in the path of travelling depressions. The successive passage of anticyclones and troughs gives variable weather conditions in New Zealand in all seasons. Prolonged periods of poor or fine weather, however, are uncommon.

(2) All airstreams that flow over New Zealand have had a long sea track. Evaporation from the ocean surface leads to the air being usually quite humid. Cooling of the air by any means gives rise to cloud formation. The oceanic environment also is responsible for relatively small diurnal and seasonal temperature variations.

(3) The mountain chain running north-east to south-west acts as a barrier to the westerlies. Across this barrier there are sharper contrasts in the climate from west to east than from north to south. There is much more rain about the west of the main ranges than in eastern areas.

Winds

Westerly surface winds prevail in all seasons and there is a general tendency for the wind to increase towards the south. The mountainous terrain causes important local wind effects. Cook Strait, the only complete break in the mountain chain, acts as a funnel producing very strong and often turbulent winds. Funnelling is also of major significance through Foveaux Strait and through the Manawatu Gorge.

In summer, sea breezes are common and often extend to at least 50 km inland. On some occasions the sea breezes extend all the way from the coast to the main mountain ranges. The mean winds at all tropospheric levels are also westerly. The wind speed generally increases with height.

The predominance of westerly winds is dependent on the location and the season. At Auckland at 900 mb (about the level of the top of the friction layer), the westerlies are most dominant in *May* and *October*, and least dominant in the mid-winter and mid-summer months. Further south at this level the predominance of westerlies is least in winter.

When the westerlies are strong, mountain waves and lee waves are frequently observed to the east of the ranges of both islands. The up-draughts and downdraughts are occasionally sufficiently strong to prevent aircraft from maintaining altitude. Severe turbulence may be experienced, particularly in the rotor zones at low levels.

Strong easterlies sometimes occur in about the lower 10 000 ft, especially when there is a depression over the north Tasman Sea. In these circumstances turbulence is often encountered west of the mountain ranges.

Thunderstorms

Places in the north and west of New Zealand report 15–20 thunderstorm days per year, but in most eastern areas, less than 5 are reported. Frontal thunderstorms occur in all seasons. Non-frontal thunderstorms are most common in summer. In some areas there is a marked diurnal variation in the incidence of thunderstorms.

Tornadoes

About 17 tornadoes are reported in New Zealand each year. They mainly occur in western areas and the Bay of Plenty. Over most of the country there are less than eight days of hail per year, but at places in the SW of the South Island, hail is reported on about 20 days per year.

22.2 North Auckland

This narrow peninsula to the north of Auckland City enjoys a comparatively warm climate in comparison to the rest of New Zealand. Fronts, which pass over this area, are usually weak, and in summer the passing of a front is often only marked by a change of wind direction.

In winter, cold fronts are often vigorous to the south of the area but to the north they are less active. This area lying in the warmer latitudes and having an 'on-shore' and moist prevailing wind is a favourite area for the formation of convective type clouds despite the low terrain. These clouds have maximum activity during the afternoon and bring frequent showers to the area.

With a depression to the NW of the area, in particular a tropical cyclone, the peninsula receives warm moist tropical air from the NE. Under these conditions the whole area is covered with low stratus cloud and rain. In the areas near the centre of the depression strong winds, often up to gale force, sweep the area bringing in heavy rain and low cloud.

The Auckland peninsula has many low-lying areas often adjacent to tidal inlets, and in these areas where the cold but moist air drains, radiation fog will often occur. The fog, although deepest just after sunrise, soon disperses as the strength of the sun increases. Occasionally the fog may lift into a thin layer of stratocumulus.

Surface temperatures:	Summer (January)	16–23°C
	Winter (July)	12–20°C
Tropopause height:	Summer	43 000 ft
	Winter	35 000 ft

22.3 The Kaikoura coast including Christchurch

This area is sheltered from the prevailing westerly winds by the Southern Alps, and the area enjoys a comparatively dry climate. The approach of a weak to moderate cold front is barely perceptible with only a small increase in wind strength. If the front is vigorous with strong winds both before and after the front, the area will be swept by a very marked Föhn wind, known locally as a 'nor'wester'. If the following wind is south-westerly, there is little change in the weather as the front passes, and after its passage the only visible change is a clearance in the medium cloud. If the wind behind the front is from the south or SE, the front moves up the coast from the south bringing an abrupt change in the weather. The front is accompanied by a line squall, with the wind rapidly backing from westerly to southerly.

The front brings heavy rain, which soon eases off, as the front moves northwards. Heavy showers are common in the cold southerly airstream, and they will persist until the wind decreases in strength or veers back to a westerly direction. When a depression is centred off the coast, the on-shore southerly winds bring low cloud, heavy rain and poor visibility.

Fogs are common, particularly in the coastal areas, and the fogs are of both the radiation and sea fog types. The fogs form rapidly and are very persistent. They are therefore a serious hazard to flying.

Surface temperatures:	Summer (January)	12–21°C
Tropopause height:	Winter (July)	02–12°C
	Summer	38 000 ft
	Winter	33 000 ft

22.4 Summary

October to March (the southern summer)

The subtropical high has a great influence on the weather in the Tasman Sea and New Zealand. It is not a static system, but consists of easterly moving anticyclones at $\cong 35°$ South, with a frequency of about five per month. Their paths are furthest south in the late summer and early autumn when they pass centrally across New Zealand. Troughs of low pressure are found between the high-pressure cells, but they are usually fairly weak in this season.

From December, there is the slight risk of a cyclone crossing the area from the north. January to March is the vulnerable period, but even then the average frequency is no more than one every 2–3 years.

Thunderstorms are not numerous over New Zealand. The frequency is greatest in the north and west where thunder is heard on 15–20 days during the season. East of the mountain ranges the figure is much less.

April to September (the southern winter)

The paths of the anticyclones are in the lower latitudes and lie furthest north in spring. The weather is controlled by a succession of lows, which mostly move SE near or to the south of the region. Each low brings its frontal weather, and the cold front in particular may be severe. Prolonged periods of bad weather, however, are unusual, because the systems rarely stagnate.

Very few thunderstorms occur over New Zealand, but there is some fog. There is a possibility of mountain waves near Christchurch. The surface winds are variable. The upper wind is westerly averaging 60–70 kt at the 200 mb level, being on the southern fringe of the subtropical jet stream.

Chapter 23

Weather in the Pacific

Weather conditions over the North Pacific have many similarities to the weather over the North Atlantic. Both areas are affected by travelling depressions along a polar front. There are large landmasses deployed to north, which include Siberia, China and Alaska. The main differences between the two oceans are:

(1) The sharper temperature contrasts result in a greater depression activity; the route concerned coincides approximately with one of the most distinct frontal zones of the world.

(2) For the same reason, the winds aloft are generally stronger than those over the North Atlantic. In particular, when the polar jet combines with the subtropical jet stream over the west North Pacific in winter, very high wind speeds result.

(3) The pronounced monsoon circulation over the west North Pacific summer and winter.

(4) The high frequency of tropical cyclones (typhoons) in the North Pacific.

23.1 Surface pressure systems and general weather

The average mean sea-level pressure distribution and position of fronts are given in Figs 23.1 and 23.2.

23.2 The main pressure systems

Aleutian low

This dominates from October to March. Pressure in the centre $\cong 1000$ mb. It appears on the mean charts as a result of the passage of a constant succession of depressions along the polar front, which lies roughly from the SW to the NE towards Alaska. The sharp temperature contrasts in winter produce stronger fronts and deeper depressions. Also, the depressions along the polar front reach much lower latitudes than in summer compared with the North Atlantic.

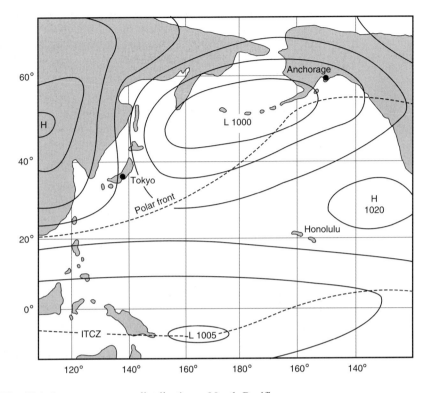

Fig. 23.1 January pressure distribution – North Pacific.

Some of the depressions form over Mongolia and Siberia and enter the ocean by way of the Sea of Japan or Sea of Okhotsk. Secondary lows often form on the southern side of these depressions and frequently become the predominant part of the system. The primary low then stagnates and degenerates over the west Aleutians or some other northern part of the ocean while the secondary moves towards the central or east Aleutians or the Gulf of Alaska. However, many form over China or the East China Sea and generally pass in the vicinity of Japan. These depressions have marked warm sectors with wide precipitation areas at the fronts. They usually deepen considerably as they cross Japan towards the Aleutians.

At the rear of these depressions very cold continental air sweeps out over the relatively warm ocean and instability clouds of particular violence may develop. The depression activity along the polar front is the immediate cause of the weather in this area. Temporary fair weather occurs in the cold migratory highs between successive depressions.

In summer, the centre of low pressure has moved further north, the pressure in the centre ≅ 1010 mb. The belt of frontal activity and disturbances is less active and is displaced towards the north. Although secondary depressions do move east across the Aleutians and Gulf of Alaska, as a rule they are not intense and move much slower than in winter.

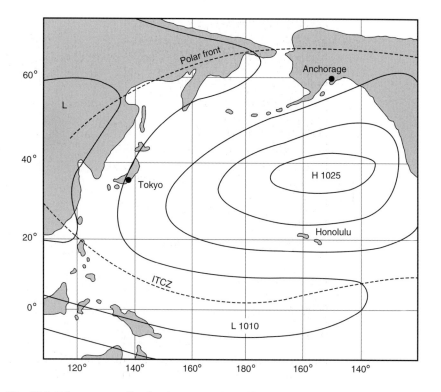

Fig. 23.2 July pressure distribution – North Pacific.

In summer, and more rarely in winter, typhoons occasionally move into higher latitudes where they take on the characteristics of an extra-tropical depression. Some of these lows affect the weather as far north as the Aleutians.

North Pacific subtropical anticyclone

This is a semi-permanent high pressure feature centred between 25° and 35° North in the east North Pacific, with a ridge spreading north into the Gulf of Alaska in the summer; pressure in the centre ≅ 1020 mb in the winter, and ≅ 1025 mb in the summer. Weather is fine and surface wind is generally westerly north of the centre.

Cold Siberian winter high (mid-October to March)

The predominant air flow over the west North Pacific is NW (or NE depending on location). A monsoon caused by this high is formed of cold polar Siberian air. This flow is present for a large part of every winter. However, there are frequent breaks in the monsoon flow, caused by depressions or cold anticyclones moving east or NE across, or near, Japan.

The monsoon brings about a remarkable contrast in weather between the two sides of Japan. The NW slopes are overcast with frequent rain or snow, while the south and east side are brilliantly sunny and clear.

Thermal low over the Asiatic continent in summer

Summer over the west North Pacific is dominated by a broad current of equatorial or tropical Pacific air towards the north. This is the SW (or south depending on location) monsoon. In mid-summer the warm, damp monsoon air covers all Japan and conditions are quiet and hot, with some thunderstorms.

Along the coasts of Japan washed by the Oya Shio current south or SE winds bring in sea fog. The northerly drift of the warm monsoon air is accompanied by disturbed conditions. The northern edge has numerous, complex, shallow depressions formed on it and these move east out into the Pacific. During June and early July the depressions move directly across Japan and bring frequent heavy rains, often prolonged. This rainy spell heralding the arrival of the southern monsoon is referred to as the *Plum rains*.

Later on in summer the track of the depressions along the leading edge of the monsoon lies further north. The cold fronts of these depressions, however, occasionally penetrate to the south and are often accompanied by heavy rain. *Typhoons become a real danger in summer and autumn.*

Upper winds and temperatures

The predominant upper air flow is westerly. The axis of the frontal depressions generally tilts NW over the cold air and the centre of low pressures aloft are found over Siberia and Kamchatka.

There is great daily variation both in direction and speed of the upper winds. An important feature is the jet stream, which is strongest and takes its most southerly position in winter.

Polar jet stream

This is associated with the polar front. The core of maximum wind velocities is found at altitudes of 30 000–40 000 ft usually vertically above the 500 mb position of the front. The jet stream roughly flows along the front, 350–700 km behind cold fronts and 700–1400 km ahead of warm fronts. It frequently flows at right angles across newly occluded fronts. Speeds in the core are usually 100–200 kt. The polar jet stream is very variable in position and strength, occurring in latitudes 30 to 60°N. Generally, the jet stream is orientated east–west, but frequently the existence of deep troughs and ridges will give it a wave-like form.

Subtropical jet stream

It is found at 20 to 35° North in winter and 40 to 50° North in summer with maximum speeds at about 40 000 ft. In summer, speeds are relatively low. In winter it enters the west Pacific via northern India, southern China, or eastern China and thence off the southern coast of Japan. Maximum speeds averaging 150 kt are found in the west Pacific.

The subtropical jet stream is a regular climatological feature. Very high wind speeds may occur occasionally over the west North Pacific in winter. In particular, when the polar jet streams combine with the subtropical jet stream, winds up to 300 kt may be encountered.

23.3 Summary

Freezing level

	Winter	Summer
60° North	Surface	6000 ft
30° North	8000 ft	16 000 ft

Tropopause height

	Winter	Summer
60° North	29 000 ft	36 000 ft
30°North	*37 000–54 000 ft	*43 000–51 000 ft

* Depends upon whether it is the polar or tropical tropopause.

Visibility

Coastal areas are prone to advection fog. Radiation fogs with clear skies and cooling surface can affect any river valleys and inland areas. Sea fogs are common at the junction of two sea currents particularly just south of Japan.

Chapter 24

Weather in North America

Central North America has the typical climate of a continental interior in middle latitudes with hot summers and cold winters, yet the weather in winter is subject to marked variability. This is determined by the steep temperature gradient between the Gulf of Mexico and the snow-covered northern plains. A further influence is the shifts of the upper wave patterns and jet stream (see Chapter 2, Section 2.3).

Unlike Europe, the West Coast is backed by Pacific coastal ranges rising to over 9000 ft (Mount Rainier in Washington State is 14 408 ft AMSL). This range effectively prevents maritime influences extending inland. Therefore, there is no extensive maritime temperature climate such as experienced in Western Europe. The climatic conditions are similar to the coastal mountains of Norway and those of Southern Chile and New Zealand under the influence of the southern westerlies. These mountain barriers cause a sharper demarcation of weather across a short distance, both horizontally and vertically.

Cyclonic activity in winter is much more pronounced over central and eastern North America than in Asia, which is dominated by the Siberian anticyclone, and consequently there is no climatic type with a winter minimum of precipitation in eastern North America.

A point to note is that there are no mountain barriers running east–west to block the very cold air from sub-polar and Arctic sources. Likewise, warm maritime air from the Gulf of Mexico can extend well to the north.

Maritime influences in eastern North America are limited considerably by the fact that the prevailing winds are westerly, so the temperature characteristic is of a continental type. Most of the moisture affecting the central part and eastern half of the United States actually comes from the Gulf of Mexico. This makes the precipitation patterns quite different to that found in eastern Asia. Little rain, however, originates from the Pacific (in the interior). The Atlantic is an additional source of precipitation at the east coast, mainly in the winter.

24.1 Main pressure systems

There is a prominent trough at the middle troposphere over eastern North America in both the summer and winter. See Fig. 3.3, 500 mb chart for March, where an upper level trough can be identified. This is most likely the result of an upper lee-trough caused by the effect of the mountain range on the westerly flows at this altitude. This wave pattern thus produced causes depressions in continental polar air to move south-eastwards over the Midwest, and depressions along the eastern seaboard move north-easterly.

The Pacific coast sees the greatest cyclonic activity occurring during the winter. Also the Great Lakes area, but the Great Plains sees the greatest number occurring in spring and early summer. The surface heating during the summer causes a quasi-permanent low-pressure area, whereas the mid-troposphere is virtually permanently affected by the subtropical high-pressure belt in the middle troposphere. The West Coast in summer is affected in the main by the Pacific anticyclone, whereas the south-eastern United States is affected by the permanent Atlantic subtropical anticyclone.

There are three main tracks of depressions during the winter.

(1)   The first group follows a near zonal path at about 45° to 50° North.
(2)   The second group curves southwards over the central United States and then turns NE towards the Gulf of St Lawrence.
(3)   The third group of depressions forms on the polar front. In winter the polar front is lying off the east coast of the United States. The depressions move towards the north-east into the general area of Newfoundland.

Some depressions originate over the Pacific. However, on crossing over the high western ranges, evidence of their presence can be seen as an upper air trough. They can then redevelop on the lee side of the mountains. This air mass is described as modified polar maritime (mPm). This phenomenon is very evident in NW Canada, where the western mountain ranges are higher than in the United States. The Arctic front is over NW Canada in winter. The frontal zone sees the meeting of mPm air from the Gulf of Alaska and very cold, dry Arctic continental or polar continental air originating from the continental source regions.

In dealing with this area, it is well to mention that Canadian meteorologists identify a third frontal zone, which is positioned between the Arctic front and polar fronts. This third frontal zone is described as the *Arctic maritime frontal zone* and is evident when Arctic maritime and polar maritime air masses meet and interact along their boundary.

From the climatological point of view, a note should be made regarding ocean temperatures, in particular, the North Pacific and North Atlantic circulation. The circulations in both oceans are similar; however, in the North Pacific, the drift from the Kuro Shio current off Japan (a warm current) is moved by the westerly winds towards the western coast of North America. The latitudes affected are $\cong$ 40° to 60° North. The sea temperatures are not as warm as the equivalent position in the North Atlantic (Gulf Stream), being some 2° to 3°C lower in temperature. Furthermore, the Alaskan coast prevents the drift further north.

The effects of the Atlantic currents on North American climatology are, however, limited. There is local moderating of minimum temperatures at the coast. The cold Labrador Current keeps temperatures low off the coast of Labrador and Newfoundland during summer, but the lower incidence of freezing temperatures in January along the same coasts is due to depressions moving into the Davis Strait, carrying Atlantic air northwards.

A note should be made about the ocean currents near Newfoundland (see also Fig. 2.3). *The Labrador Current is a cold current. It flows southwards* from Baffin Bay, through the Davis Strait, and south-westwards along the Labrador and Newfoundland coasts. The *Gulf Stream is a relatively warm, swift current, flowing northwards.* It originates north of Grand Bahama Island. It tracks north initially and then sweeps across the North Atlantic. It is also called the North Atlantic drift.

The significance of these two ocean currents is that this is the mechanism to form advection fog over the Newfoundland fishing banks. Warm, moist air flowing off the warm Gulf Stream over the cold Labrador Current can form advection fog. This can take place at any season of the year, and being an advection process can form any time of day or night.

Winter (December to February)

This is the worst period of the year for flying conditions across most of the continent. Exceptions are (a) the east slope of the Rocky Mountains (where March or April is the worst period) and (b) coastal terminals in the east (Boston and New York) and north-west (Seattle). At these eastern terminals, low cloud and poor visibility are most frequent in May and at Seattle in late autumn.

In the east, depressions bring widespread snow, sleet, rain, freezing rain and fog. Surface winds are mainly easterly when the weather is bad, and winds from between west and north bring clearer conditions. South to SW winds sometimes bring fog, especially with a long track from the Gulf of Mexico, when the Great Plains, almost as far as the Rockies, may also be affected.

In early winter (November) the land around the Hudson Bay area is snow covered; with a westerly wind the air is warmed over the open water by $\cong 11°C$ or so. The air stream becomes moist and produces heavy snowfalls along the eastern and southern coasts. In January, Hudson Bay is frozen, and no effects are noticed.

The Great Lakes create similar effects. Heavy snowfalls take place on the southern and eastern shores with northerly or north-westerly winds. A point to bear in mind is that the lakes retain heat into the early winter, and polar continental or Arctic continental air is warmed and takes up additional water vapour, hence the snow falls on the easterly and southern shores, but also there is orographic displacement, enhancing snowfall. Because the lakes are a heat source, a low pressure trough forms, and convergence into the trough or troughs exacerbates the precipitation.

There are two types of synoptic situations which are of particular importance in controlling temperature in North America.

1 *Cold spell*. These are the result of outbreaks of cold polar continental (Pc) air. During the winter these cold outbreaks can penetrate deep into central and eastern North America; they can even reach as far as the Gulf of Mexico. A cold wave is locally defined as a drop in temperature of at least 20°F (11°C) in a 24-hour period over most of the United States.

The synoptic situation associated with these cold spells is where an anticyclone builds up behind a cold front. They are aligned $\cong$ north–south. The weather is then clear and dry, with cold but strong winds. If, however, snow is deposited during the passage of the cold front, then the following winds will create blizzard conditions. This is a common phenomenon over the northern plains.

2 *Warm spell*. This temperature fluctuation is caused by the *Chinook* winds on the lee side of the Rockies. These winds are Föhn effect winds. The air mass originated over the Pacific and has passed over the Rocky Mountains barrier and descended as a warm dry wind on the lee side. The onset of the Chinook can bring in sudden temperature increases. Should there be any snow on the ground, it can be thawed very rapidly. The word 'Chinook' is a Native American word meaning 'snow eater'. A temperature increase in the order of 22°C in 5 minutes has been observed! Chinook winds are common during December to February. The effects of the warm winds are noticed up to at least 50 km from the foothills in Colorado.

In Canada, the term 'Chinook effect' is used to describe the same phenomenon, but as the mountain range is higher, effects have been noticed considerably further inland in areas such as south-west Saskatchewan.

In Fig. 24.1, the pressure systems are shown for January, and also the dominant winds. The Pacific low pressure (Aleutians low) and Atlantic low pressure (Icelandic low) are the permanent low pressure belt. Also, the Pacific polar front (1) and the Atlantic polar front (2) can be seen. The Rocky Mountains range acts as a barrier to the tropical maritime air over the Pacific. The on-shore airflow from the Pacific gives thick cloud and rain along the West Coast; also, the on-shore airflow from the Gulf of Mexico gives rain. The interior of the continent is dry and very cold in the north. The Appalachians sees very wet conditions on the east side, and drier conditions on the west side.

Most occasions of low ceiling and visibility occur with a warm front to the south over the Carolinas or near the coast. Winter warm fronts nearly always slow down near New York and Boston, and may stop near Long Island. The bad weather is then persistent.

Depressions near or north of the Canadian border seldom bring more than

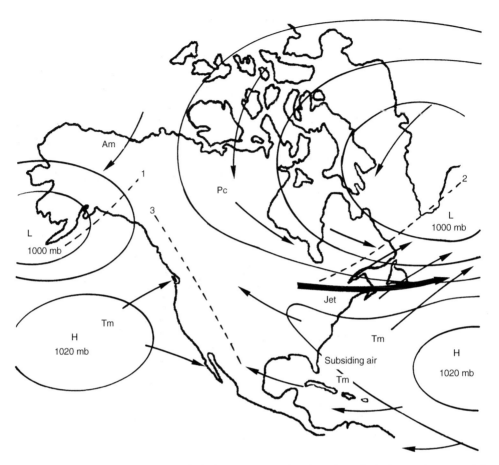

Fig. 24.1 North America, pressure and wind – January. Polar air and tropical air can alternate, giving rapid extreme temperature changes. 1 – Polar front Pm or Am and Tm. 2 – Cold frontal zone Pc and Tm. 3 – Frontal zone on the lee side of the Rockies. Considerable modification to Tm airmass after descent.

brief periods of precipitation and low ceiling to the region of the Great Lakes. Those which pass far to the south are generally associated with freezing temperatures over the Mid-west but with fine flying conditions. All eastern airports are liable to be closed by deep snowdrifts when intense lows develop near Cape Hatteras, North Carolina.

Over the continental plateau (east of the Rocky Mountains) intense storms are infrequent, but orographic effects help to produce considerable precipitation and low ceiling at times. The most extensive and most persistent bad weather type is radiation fog associated with the plateau high. On the other hand, flying conditions in the high may be excellent. The chances of fog are greatest soon after rain or wet snow. Las Vegas is almost entirely free of low cloud, and provides a reliable alternative when the West Coast airports are closed. During periods of strong winds aloft mountain waves are common to the lee of high mountains. They are very marked near Reno and Denver.

The West Coast is exposed to depressions from the Pacific, see Fig. 24.2 in which the main *Pacific depressions* are shown. The Aleutians low pressure area (1) sees many Pacific low pressure systems becoming stationary in this area in any season, but mainly winter. Occlusions can cross the Rockies if the airflow is westerly. Chinook wind is likely to be warm. The Californian depression (2) is rarer, as even in the winter a high-pressure area (part of the permanent high-pressure belt) is in this position. However, they do appear, and when they do (winter only) they are likely to be fast moving and severe along the whole track across to the east of the mainland. Frequency has been observed to increase during an El Niño year.

Fig. 24.2 North America – Pacific depressions.

From Fig. 24.3, the position of *lee depressions* is shown. The depression over Alberta (3) can appear in any season, but is frequent and intense in winter. The depression over Colorado (4) can also appear in any season. A large air mass contrast is likely as it moves NE and develops; it moves quite fast, and can be intense. The cold front can be very active and can create thunderstorms and tornadoes.

Fig. 24.3 North America – lee depressions. These depressions can form and develop independently.

The most violent weather occurs when a series of lows forms close to the coast on the trailing edge of a cold front. Coastal airports may then be closed for some hours, and even when the cloud base lifts above limits the surface winds are strong enough to seriously affect take-off and landing. The San Francisco Bay area is partially protected by coastal hills during active storms, when more exposed stations (e.g. Los Angeles) have lower ceiling and visibility. Nevertheless, poor airport conditions are more likely to be associated with radiation cooling, and smog in the Los Angeles Basin may be persistent. A very marked temperature fluctuation associated with the Chinook wind should be remembered (Fig. 24.2).

An anticyclone is likely to form well to the north in polar continental air behind an active cold front. It can move fairly quickly south, then SE or east; it may merge with, and reinforce, the Azores high, and can bring intense cold well south in winter. It is pleasantly cool in summer and clears smoke, etc., giving excellent visibility (see Fig. 24.4). A note should be made about the high-pressure region during the winter. This area is intensified by the very cold surface. It is a cold anticyclone, and does not extend into the upper tropo-

Fig. 24.4 North America – anticyclone.

sphere. High pressure at the surface with a cold column of air sees the high pressure weakening with height and becoming a low-pressure zone aloft (see Chapter 3, 'Upper Winds and Jet Streams'). On surface synoptic charts for this time of year (winter) the North American high-pressure area can be seen to join up with the Azores high-pressure area, the pressure being ≅ 1020 mb (see Fig. 1.6). The Azores high is formed by descending air on the poleward limb of the Hadley cell, and is a warm anticyclone. Although some surface charts show these high-pressure areas joined and extending over both the sea and the landmass, they are quite different in their formation and upper level characteristics.

Summer (June to August)

In late June, there is a rapid northward displacement of the subtropical high-pressure areas in the northern hemisphere. In North America, this has the effect of moving the depression tracks further northwards. This results in turn in less precipitation over the northern Great Plains. However, the south-westerly anticyclonic airflow that affected areas like Arizona up to late June is now replaced by moist air from the Gulf of California, and this is the cause of the onset of the summer rains.

Frontal systems are weaker and the frequency of lows that form over the southern states is much less. Significant weather is mainly local stratus or fog between midnight and dawn, mainly on the north side of weak fronts lying east–west. Often, however, they form the baseline for the development of extensive thunderstorms.

Hudson Bay in the summer sees the water temperatures remaining cool, about 7° to 9°C, and this keeps temperatures down along the shorelines, particularly the eastern shores. A similar situation exists on shorelines at the Great Lakes.

Along the New England coast fog is likely with easterly winds, especially in July. This is the period, and in particular July, of major thunderstorm activity. Near the east coast, severe storms are most likely over the Alleghenies, and near the Great Lakes they tend to be most frequent SE of Lake Erie. Close to Lake Michigan and over the Great Plains, thunderstorms occur at night. The frequency of thunderstorms increases westwards over the Great Plains to a maximum of 10–15 days in July over the Rockies.

A note should be made about night-time thunderstorms. The associated precipitation accounts for over 60% of the summer precipitation in Nebraska, Iowa and Kansas. It takes place mainly between 1800 and 0600 hours. It is most likely that the vast area of the Great Plains produces a large-scale circulation east of the Rocky Mountains. This has a tendency to cause low-level divergence, and mid-level subsidence during the day. This will inhibit vertical thermal development; however, at night, convergence takes place and this results in rising air, and instability forming thunderstorms. This is an unusual sequence of events over a continental landmass.

A further significant observation is that $\cong 100°$ West, night-time southerly jet streams have been recorded. They are low-level phenomena situated between $\cong 1600$ and 3500 ft. These winds are most likely related to the large-scale inversion over the mountains, and they are also probably responsible for the influx of moisture and convergence.

This season is also notable for tornadoes and hailstorms. The former are most likely in the states of Iowa and Kansas and the latter along the east slope of the Rockies.

In Fig. 24.5 it can be seen that low pressure dominates the northern regions. The Atlantic depressions are further north than in winter. The West Coast is cool with thick dense cloud in the north, and some rain. The interior of the continent is dry, and hot in the south. The Appalachians has rain both sides.

The frequency of thunderstorms decreases markedly to the west of the continental plateau, and the Californian valley enjoys good conditions. Along the coast, however, July is the worst month for sea fog at exposed stations like Los Angeles, but just inland it becomes stratus. At San Francisco it is mainly stratus with good visibility below.

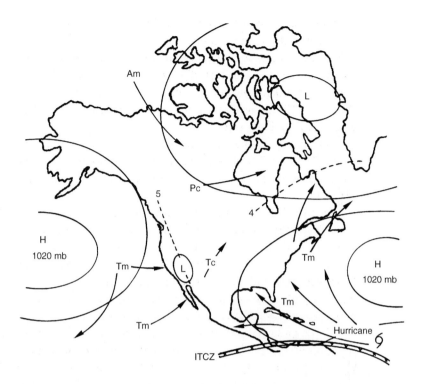

Fig. 24.5 North America, pressure and winds – July. Atlantic frontal zone (4) further north than in winter; Rocky Mountain frontal zone (5) weaker than in winter.

A cold current (the Labrador Current) influences the Labrador coast. This sea current is analogous to the Oya Shio current off Eastern Asia. In both cases, the climatic significance is limited by the prevailing westerly winds. This current moves drift ice in its track off the coast, including the Newfoundland coast until at least June. This keeps the coastal temperatures down in summer.

One significant effect of the Labrador Current is the formation of advection fogs, which are very frequent between *May and August* over the Newfoundland fishing banks. The meeting of the two sea currents, the Labrador Current and the Gulf Stream brings about these fogs. Warm moist southerly air flowing from the Gulf Stream over the cold Labrador Current is rapidly cooled. The fogs may persist for days. The wind speeds vary from light to very strong. At Cape Race (Newfoundland) there are over 150 days a year when fog is present; the peak month is August.

Spring (March to May) and autumn (September to November)

In addition to the changes typical of transitional seasons, a number of special features occur in these two periods. Early spring is characterised by persistent easterlies along the eastern slope of the Rockies with dense cloud and rain.

During March and April there is a marked decrease of precipitation in California, due to the extension of the Pacific high-pressure area. However, precipitation increases in the Midwest because of the greater cyclonic activity and warm tropical maritime air coming from the Gulf of Mexico.

During May cold fronts tend to become stationary just off the east coast between Nantucket and Norfolk, with low cloud, rain and drizzle east of the Alleghenies for several days at a time, especially north of Philadelphia. Coastal stations like Boston and New York are particularly affected, but just inland (e.g. Baltimore and Philadelphia), low cloud and poor visibility is still more frequent in winter. Similar conditions occasionally occur just south of the Great Lakes when fronts lie parallel to the south of the airway.

In early spring all thunderstorms are of the frontal type. Line squalls are most frequent in May and June, making en-route weather more adverse in some respects than that of winter.

Sea fog persists along the Californian coast in September, and there is a marked increase in low cloud, rain and fog further north (e.g. Seattle) as frontal depressions move across the area from west to NW. This is the month when the risk of hurricanes is greatest.

The tracks of Atlantic depressions are shown in Fig. 24.6. They form well south (Florida area) on the Atlantic polar front and move NE; they are rare in summer, and move out over the Atlantic. Hurricanes form from July to October, an average of 5 per year. They form east of the Caribbean, move west to NW, then 'recurve' to move north or NE, causing severe damage over islands and coasts. They usually die out quickly inland (see Chapter 25, 'Weather in the Caribbean').

Freezing level and icing
In July the average height of the 0°C isotherm is 15 000 ft. In January, however, the 0°C isotherm is on or close to the ground over most areas and icing problems also occur at airports, especially in the east. Near the West Coast, it rises to 4000 ft at Seattle and 9000 ft at Los Angeles.

Upper winds
On average the upper westerlies are strongest in the east and decrease westwards, but they vary considerably from day to day. During winter there is a high frequency of jet streams.

The tropopause

January
The average height of the polar tropopause is 37 000 ft between New York and San Francisco, but the actual height may fall well below 30 000 ft locally, especially when the core of a jet stream is well to the south.

Fig. 24.6 North Atlantic – Atlantic Coast depressions: hurricanes (6), polar front (7).

July
The polar tropopause is often absent at this time of year over most of the route New York to San Francisco, and its average height east of Chicago is 41 000 ft.

24.2 The importance of North American tornadoes

With the rapid development of a tropical cumulonimbus, violent eddies are set up within or near the cloud. These are mainly due to the intense shear between the updraughts and downdraughts in the cloud, the eddies forming a funnel-like structure which, with continued development, may grow downwards to the ground. The eddies in the cloud vary from 50 to 250 m in diameter (some very intense tornadoes are greater) and are usually of such intensity as to become a *very serious risk* to aircraft operating in their vicinity. If the vortex does not grow downwards, the danger to the ground is not great, but if conditions are favourable, a path of between 50 and 300 m can be ploughed through the countryside, inflicting total destruction on buildings, trees and crops.

The height of the tornado season is April to May. They form over the great lowland areas of the central and upper Mississippi and 'Tornado Alley' is the name given to those areas most affected, although tornadoes have occurred in almost every state. Texas, Oklahoma, Kansas, Iowa and Alabama seem to be the favoured states. Over 1000 tornadoes are reported every year.

Tornadoes result from the development of severe thunderstorms. The instability being a consequence of the meeting of two air masses, one a warm humid air mass origination from the Gulf of Mexico, and a dryer air mass from the north, having become dry from a long continental track. Or a dry air mass from the NW, again becoming dry from the descending Föhn effect on the lee side of the Rockies. The leading edge being described as the 'dry line'. The warm humid air undercuts the drier air, and daytime warming of the land surface provides the trigger action to make the air ascend. This mechanism can be seen in West Africa during the Line Squall Season (see pp. 35–145).

The ascending air has a steep lapse rate, and the instability is enhanced when the rising air penetrates into the drier (and colder) air aloft, the rising air can accelerate. The resulting thunderstorms can be massive, but it takes further special mechanisms to be in place to form tornadoes. This is an ongoing area of investigation by the US meteorologists, and indeed the unique band of 'storm chasers'.

Professor Fujita (Professor of Meteorology, Chicago University), has produced a tornado intensity scale. It is based in the main on the damage sustained by the passage of a tornado. It ranges from F0 to F5:

F0 Intensity, branches would be seen breaking off trees, winds less than 180 kph.
F1 Wind speed from 180 kph (this is already in the hurricane speed range).
F2 Wind speed from 250 kph.
F3 Wind speed from 330 kph.
F4 Wind speed from 420 kph.
F5 Wind speed from 510 kph.

Near Birmingham (Alabama) in September 1998, a tornado officially rated as an F5 caused such considerable damage and destruction that unofficially it may have exceeded the Fujita scale, winds over 510 kph ($\cong$ 270 kt)!

Storm chasers equipped with modern Doppler radars are improving our knowledge of the mechanisms of these catastrophic systems.

Chapter 25
Weather in the Caribbean

The weather in this area is normally under the influence of the North Atlantic subtropical anticyclone, and the NE trade-wind circulation. Also the Equatorial trough. The normal trade-wind Cu gradually builds up to greater heights as the Equatorial trough (ITCZ) is approached. Although the trade-wind circulation is very constant, disturbances occur and produce some heavy Cb and thunderstorms.

25.1 Hurricanes

This region is affected by hurricanes which form in the principal areas near the Cape Verde Islands and the Caribbean Sea (mainly off the coasts of Honduras and the Yucatan Peninsular). From observations, distinct stages can be identified, which may be described in a similar way to the formation of thunderstorms.

Formative stage

This begins when an organised circulation develops in the *Equatorial trough* or in an *easterly wave*, and ends when the disturbance reaches hurricane intensity. Deepening can be a slow progress; it can also be very rapid, and produces the characteristic 'eye' within 12 hours or so.

Immature stage

During this stage deepening continues and maximum intensity is reached. Hurricane force winds are found within a 35–55 km radius of the eye. In this stage the hurricane is almost symmetrical and covers a relatively small area; winds of hurricane force form a tight band around the centre, and cloud patterns change from disorganised squalls to narrow bands, spiralling inward to the solid wall of cloud around the eye.

Mature stage

The isobars expand without further deepening and the actual intensity often decreases, although the area covered by gale and hurricane force winds is largest. The average radius of hurricane force winds is 80–100 km with gales up to a radius of 350 km. The symmetry is lost as the area of gales and bad weather extends further to the right of the direction of motion rather than to the left.

Decay stage

The hurricane either moves inland and dissipates or recurves northwards and assumes extra-tropical characteristics.

Upper air effects

Both the size and intensity of hurricanes decreases with altitude. At 30 000 ft, the wind speeds have decreased considerably, and above 40 000 ft the area of cyclonic circulation is very small.

Frequency of hurricanes

The main hurricane season is from *June to November*, with the highest frequency from August to October. On average 7 occur per year. In May and June hurricanes originate mostly in the west Caribbean or the Gulf of Mexico. The July hurricanes most frequently form in the east Caribbean or north-west Atlantic.

   During August and September, the main source region is the eastern part of the North Atlantic. Towards the end of the hurricane season, most develop in the west Caribbean or Gulf of Mexico.

Tracks

See Chapter 6, Fig. 6.5. The southern edge of the hurricane belt is 10° North. The principal tracks are approximately parabolic, with the initial track westerly. The point of greatest curvature is seen as the track passes through north. This point moves progressively eastwards throughout the season. However, individual storms may deviate considerably from the general pattern; the average speed of travel is 10–16 kt, but this increases with track curvature.

25.2 Seasons

The Caribbean area and the adjacent Atlantic to the north and NE have two main seasons, a dry season and a wet season.

Dry season

This is from *November to April*. The ITCZ lies well to the south and the Azores high dominates, giving a high frequency of fair weather Cu reaching maximum development over the islands by day and the sea and windward coast just before dawn. From time to time, however, cold fronts and upper troughs or waves in the westerlies enter the area from the NW via the Gulf of Mexico. These polar front phenomena may even reach the northern part of South America. By that time, however, they are no longer recognisable as fronts, but have developed into shear lines or troughs, and are indicated as such on weather charts. The squalls and thunderstorms along these disturbances are much less intense than in summer, with the worst weather in the trough ahead of the front. When active, such troughs give the worst winter weather in the Caribbean.

Such systems generally become stationary near Puerto Rico, though some have been known to pass over the whole of the West Indies. Others may reverse their direction of travel and move back over the Bahamas, bringing thick layers of medium cloud and prolonged periods of slight or moderate rain.

A second type of disturbance enters the region from the east or NE and accounts for most of the rainfall in the eastern Caribbean, e.g. Barbados. These are the remnants of polar fronts that have circulated around the Azores anticyclone. Generally they are not very intense.

Rainy season

This is during the summer from *May to November*, with the wettest season from May to June. The Azores high has retreated north and east allowing the development of air mass type Cu and Cb giving thunderstorms and heavy showers. Maximum development is over the land during the afternoon and over the sea at night, but along windward coasts there is little diurnal variation. Over the SE states of the USA and many islands in the northern part of this area, Cb are almost a daily occurrence. In the south, the frequency of thunderstorms is less and along the coast of Venezuela and Guyana it becomes very small.

On average the ITCZ is close to Guyana in July, but its precise position is uncertain. Two or three rain belts are likely to affect Trinidad at the onset of the rainy season. Visibility is generally good at this time of the year. In heavy showers it is reduced temporarily, perhaps to a few hundred metres, but good visibility generally returns within the hour. Fog hardly ever occurs in the West Indies. Nassau and Port of Spain are very occasionally affected during the early morning.

The coast of Guyana has a rather different regime to the above. At Georgetown there are the two rainfall maxima typical of equatorial regions, and December is one of the wettest months, followed by a rapid decrease early in the year. Port of Spain also has a second maximum in December but it is very slight. Rainfall is often heavy but thunderstorms are rare.

25.3 Trade-winds

The trades cover those parts of the ocean between the subtropical high-pressure belts and the Equatorial trough. Hence north of the Equator they generally blow from the NE, and south of the Equator from the SE. When the ITCZ is located relatively far from the Equator, the portion towards the Equator is occupied by the equatorial westerlies, i.e. the deflected trades of the other hemisphere. This accounts for the SW monsoons of West Africa and of Ecuador and Columbia during the northern summer.

The trades become unstable at the lower levels while flowing equatorwards over the sea. The presence of the trade wind inversion, however, restricts the tops of the trade Cu to about 10 000 ft. The constancy of the trades is exceeded in no other region on Earth, but interruptions do occur. The main disturbances are easterly waves and hurricanes. For more details on these items, see Chapter 4, 'Easterly Waves' and Chapter 6, 'Tropical Storms'.

Chapter 26

Weather in South America and the South Atlantic

The South American continent extends from latitude $\cong 10°$ North to $55°$ South and displays a wide variety of climates. The climatic zones extend from equatorial on the north, through subtropical high pressure, warm and cool temperate, just reaching polar tundra climate from $\cong 50°$ South. The mountainous regions of the Andes extend almost the entire length of the western seaboard, and form a very important meteorological barrier, preventing air from the Pacific and Atlantic regions being interchanged except at high levels.

From latitude $38°$ South to $5°$ North, this barrier is effectively more than 10 000 ft in height with the exception of a gap in the region of $5°$ South (in the neighbourhood of Guayaquil in Ecuador). The highest peaks in the range exceed 20 000 ft and there are other topographic features which have significant meteorological importance. Some of these are the high ground of SE of Brazil, which exceeds 3000 ft over wide areas, and the high ground in the south of Guyana and Venezuela which separates the Amazon basin from the coast. See Fig. 26.1 for the geography of South America.

In dealing with pressure distribution and air-mass flows in temperate latitudes in the *southern hemisphere*, do not forget that circulations are opposite to those of the northern hemisphere. Cyclonic flow is clockwise, and anticyclonic flow is anticlockwise, i.e. the passage of a front will see wind backing, and not veering as in the northern hemisphere (see Fig. 26.2).

The distribution of sea-surface temperature in the oceans bordering South America is also of major importance in determining the climates of the continent (see Chapter 2, Section 2.4 and Fig. 2.3). The east coast of South America from $\cong 10°$ South is influenced by the Brazil current, a warm current flowing southwards. The west coast from $\cong 30°$ South is influenced by the Peru current, a cold current flowing north to NW.

Due to the position of the southern reaches of the continent, a cold sea current sweeps around Cape Horn from the west (the west wind drift) and flows northwards off the east coast of Argentina. This north-flowing current is called the Falkland current and brings cold water north as far as latitude $35°$ South. Along the east coast further to the north, the Brazil current flowing southwards brings water warm for its latitude to the east coast of Brazil,

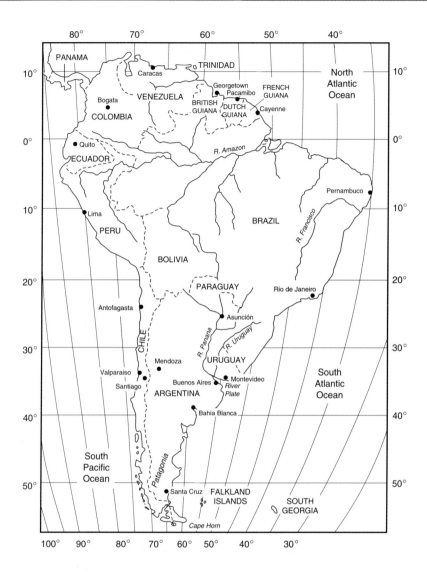

Fig. 26.1 Geographic map of South America.

meeting the Falkland current about latitude 35° South and passing to the east of it (Fig. 26.3).

This results in a strong concentration of isotherms of sea temperature along a line extending in a south to SE direction from the region of the River Plate. The gradient of sea temperature in this region, although not so great as that found off the east coast of North America, is undoubtedly of importance in the maintenance of the polar fronts along which wave depressions move SE from northern Argentina and Uruguay.

This marked east–west gradient of sea temperature also has an important

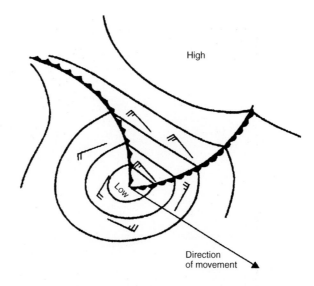

Fig. 26.2 A warm sector depression in the southern hemisphere.

effect on some easterly air streams arriving on the coast of Uruguay and Argentina. Although originally of polar origin such air currents have to pass from relatively warm to cooler water before reaching the coast. They therefore become moist and stable in the lower layers and may give rise to fog or low stratus.

In the Atlantic Ocean to the south-east of South America lies another concentration of isotherms of sea temperature extending along a line in an east to NE direction from a point approximately 60° South, 70° West to a point 50° South, 30° West. This is a region of *convergence* of relatively warm, surface water moving from the north with cold water moving from the south, both of which streams rotate round the south polar regions in the Antarctic circumpolar current. This line of convergence is known as the *Antarctic convergence*, and is often marked by a sharp contrast of sea surface temperature. The alignment is due to the deflection of the Antarctic circumpolar current towards the north in this region, as a result of the presence of the Graham Islands and certain submerged ridges.

The gradient of surface temperature at the Antarctic convergence, together with that at the boundary of the pack ice, which lies to the south in a generally zonal direction, is of major importance in maintaining the *Antarctic front*. Depressions forming in this region frequently take on an east to NE track along the general alignment of the surface isotherms (see Figs 26.4 and 26.5).

Figures 26.4 and 26.5 show the mean pressure distribution for January and July. The subtropical anticyclones over the oceans lie at about 30° South, and the Equatorial low-pressure trough lies in the region of the Equator. There is a slight northerly displacement of the subtropical anticyclones and equatorial

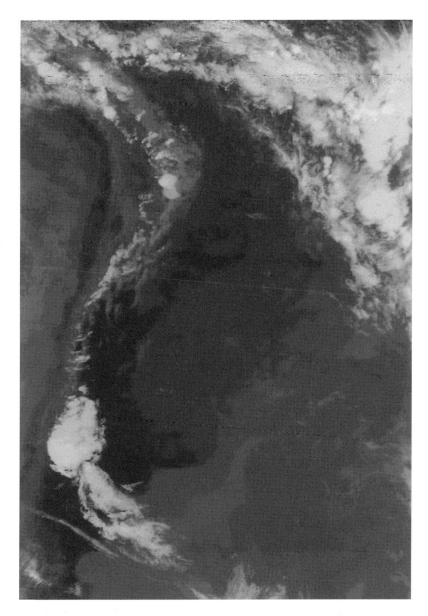

Fig. 26.3 IR Image from NOAA 12. Note the mountain barrier effect on air mass and consequent cloud formation along the line of the Andes. The Falklands current can be seen as a lighter (colder) tongue parallel with the Argentinian coast. Note too there is a jet stream located south of the cloud mass over southern Argentina – the 'thin' line of cirrus cloud outlines the axis. (This image was supplied by John Coppens, University of Cordoba, and is reproduced with the permission of the University, and acknowledgements to NOAA.)

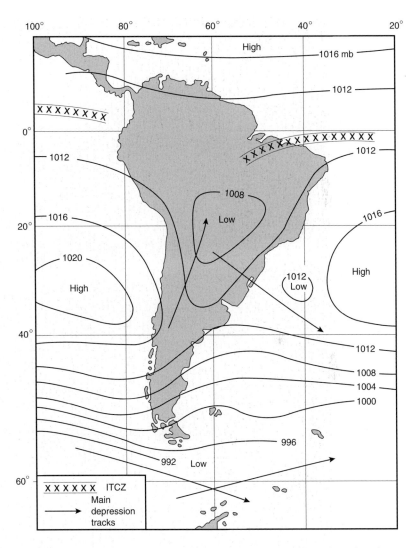

Fig. 26.4 The mean pressure distribution in January (summer), and tracks of depressions.

low-pressure belt in July, and a slight southerly displacement in January. In January (summer) a continental low-pressure area develops as shown in Fig. 26.4, and it is here the highest temperatures occur. However, there is no corresponding continental anticyclone in winter because the continent is not sufficiently wide in temperate and subtropical latitudes for its cooling to give rise to a stagnant cold air mass and consequent high pressure.

It is observed that even in July the mean pressure chart shows a trough over the South American continent between the subtropical anticyclones over the oceans, this being due to the tendency of frontal troughs to become slow moving over northern Argentina.

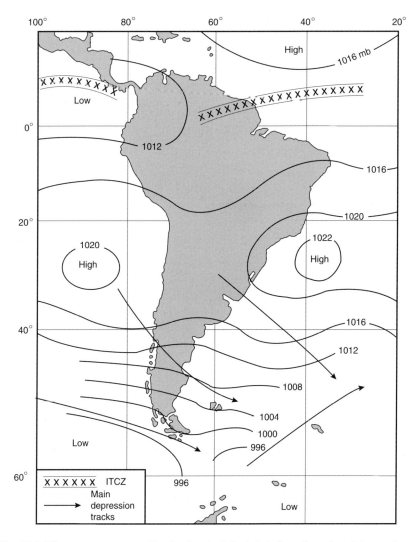

Fig. 26.5 The mean pressures distribution in July (winter), and tracks of depressions.

The main tracks of depressions are indicated in Figs 26.4 and 26.5. With the exception of those on a north/NE track to the east of the Andes in summer (January), these depressions normally have a frontal structure similar to that observed in temperate latitudes of the northern hemisphere.

The depressions which move on a north/NE track to the east of the Andes are shallow 'heat lows' produced by the heating and consequent reduction in density of the relatively stagnant air over the continent. The depressions which move from NW to SE in the region of Cape Horn probably form on a polar front which develops in the region about longitude 110°–120° West between two cells of the South Pacific subtropical high-pressure belt. They are usually occluded by the time they reach South America. In winter, similar depressions also form further to the NE and approach central Chile crossing the Andes in latitude 40°–45° South.

The ITCZ over the Atlantic remains north of the Equator throughout the year. Over the landmasses it sinks southwards in sympathy with the developed equatorial seasonal low in the southern hemisphere summer (January) but retreats northwards by July. Note that there are no tropical cyclones in the South Atlantic due to an absence of the ITCZ and the prevailing low sea temperatures. The ITCZ in fact does not move south of 5° South, so no Coriolis force is available to help initiate a vortex.

26.1 Air masses of South America

The following are the main air masses which affect the climate of South America.

(1) Antarctic (A)
(2) Polar (P), which can be subdivided into polar maritime (Atlantic) (Pm(A)), polar maritime (Pacific) (Pm(P)) and polar continental (Pc)
(3) Tropical (T), which can be subdivided into tropical maritime (Atlantic) (Tm(A)), tropical maritime (Pacific) (Tm(P)) and tropical continental (Tc)
(4) Equatorial (E)

The continent of South America is too narrow in temperate latitudes to give rise to a true polar continental air mass. However, polar maritime (Pacific) air, which crosses the Andes between latitudes 40° and 50° South, and then turns NE without crossing the Atlantic Ocean, reaches northern Argentina with many of the characteristics of a polar continental air mass and is thus classified as such.

Antarctic air

The source of Antarctic air is the ice cap of the Antarctic continent and the stretches of sea ice which surround it. The region of the Antarctic continent is probably one in which subsidence of air predominates and in which Antarctic air at its source is maintained very cold and stable by the radiation loss of heat. On passing northwards into the circulation of the polar depressions, Antarctic air is rapidly warmed and moistened from the surface upwards by the relatively warm sea surface over which it passes. It thus reaches the latitude of extreme South America as a cold, moist, unstable air mass. Temperatures are in the order of 4°C and relative humidities of over 90% at the coast. However, the absolute humidity is very low at these low temperatures.

Antarctic air rarely spreads along the west coast of South America beyond latitude 50° or 45° South without having such a long sea track as to have been modified into polar maritime air. However, particularly in winter, a small anticyclone or ridge of high pressure frequently builds up behind a series of depressions, and a southerly air stream may extend up the east coast directly from the region of the Antarctic continent to latitude 35° or 30° South carrying Antarctic air as far north as Uruguay and northern Argentina.

Polar air

The source of polar air in the southern hemisphere is the broad westerly air current of the southern oceans. Air entering the region of the westerlies from the Antarctic or from the subtropical anticyclones and having a sufficiently long eastward track parallel to the isotherms of sea temperature acquires characteristics typical of a polar maritime air mass.

The westerly polar flow across the Pacific on reaching the South American continent south of 40° South, takes on one of three preferred tracks, in all cases being modified by the temperature changes at the surface. One track is towards the north of the coast of Chile and it is convenient to classify it as polar maritime (Pacific), Pm(P).

As subsidence normally takes place off the west coast of South America under the influence of the subtropical anticyclone of the Pacific, the polar maritime air flowing northwards in this stream becomes warmed aloft as well as at the surface. When it reaches $\cong$ latitude 35° South it is normally characterised by an inversion of temperature in the region of 2000 to 5000 ft, below which the air is moist and has a steep lapse rate. Above the inversion the air is dry. A layer of cloud is frequently present below the inversion.

However, on occasions in winter and spring, when the Pacific anticyclone is displaced westwards or northwards, polar maritime air can move along the west coast undergoing little or no subsidence. It then reaches the latitude of Valparaiso with a steep lapse rate and gives rise to frequent showers and local thunder. Therefore, the resulting weather is dependent on the stability of the air mass and this in turn is dependent on the position of the Pacific anticyclone.

Because of the barrier of the Andes, polar maritime (Pacific) air rarely appears as a surface air mass east of the Andes. However, a second track of Pm(P) air on some occasions in the winter when it is deep and unstable crosses the Andes at high levels and enters into the circulation over northern Argentina at levels above 10 000 ft. On a few of these occasions, mainly in spring, when the air over Argentina is hot, polar maritime (Pacific) air sinks to the level of the plains after crossing through the Andes passes, and gives rise to a hot, dry, Föhn wind known locally as the '*zonda*'. The zonda is accompanied by clear skies and although unpleasant on the ground because of its dryness and heat, it is not of great consequence to aviation, except where it reduces visibility by raising dust. It is usually followed by a strong *pampero*, which may also reduce visibility because of the dust raised over western Argentina. The poor visibility may last some time.

The third track of polar maritime (Pacific) air masses of the westerly belt frequently turns north-eastwards and later north-westwards after crossing the southern part of South America in latitude $\cong 55°$ to $45°$ South. This air mass now approaches the coasts of Argentina, Uruguay and South Brazil from a direction between south and east having tracked virtually 2000 miles over the Atlantic Ocean, during which it has undergone considerable subsidence under the influence of the ridge of high pressure around which it was deflected. The air mass has been moistened and warmed by the sea, the surface of which becomes generally warmer along its track. The air mass now has a new designation, namely polar maritime (Atlantic), Pm(A) air.

The stability of polar maritime (Atlantic) air when it returns to the east coast of South America thus depends on the proximity of the South Atlantic anticyclone. The anticyclone can cause subsidence and warming from below. In general this is sufficient to prevent the formation of large cumulus clouds and widespread showers, but small cumulus and stratocumulus spreading out beneath an inversion of temperature are common. Coastal showers do, however, occur at times in polar maritime (Atlantic) air, and on spring afternoons convectional showers sometimes occur inland in this air mass.

Polar maritime (Atlantic) air is moist in the lower layers, and in autumn and winter, when it has spread inland over northern Argentina, radiation fog and low stratus cloud will form frequently at night. Polar maritime (Atlantic) air which has tracked well to the east may have had to cross cooler coastal waters before reaching the coast and may give rise to low coastal stratus cloud and fog on a few occasions in the year.

Polar continental air

There is no true source region for polar continental air within the southern hemisphere. However, a frequent track followed by polar air from the Pacific westerlies crosses the southern Andes south of latitude $40°$ South and then proceeds NE across Patagonia into northern Argentina without passing over the Atlantic Ocean.

Air on such a track is dried by descent as it crosses the southern Andes, causing notable Föhn effects, and is also subject to subsidence as it is deflected NE around a ridge of high pressure. It absorbs little or no moisture from the underlying ground because of the low rainfall of Patagonia, and reaches northern Argentina, after considerable warming in summer and slight warming in winter, as an exceedingly dry and moderately stable air mass, practically free of low cloud. It has thus acquired all the characteristics normally associated with a polar continental air mass in the northern hemisphere, and is conveniently classified as such.

Both polar maritime (Atlantic) and polar continental air frequently move further north before losing their identity. In winter polar air with a more or less direct track from the south will frequently pass latitude 25° South and it is not uncommon for vigorous cold fronts to reach latitude 15° South in central Brazil, causing a marked fall of temperature known locally as '*friagem*'. Shallow layers of cold air moving on this track sometimes cross the Matto Grosso, and have been identified in the Amazon basin. In summer, however, true polar air does not usually pass north of latitude 25° South on the eastern side of the Andes.

On reaching the latitude of northern Argentina, polar air masses frequently become stagnant under the slight gradients of pressure which are common in this area. Over land, these air masses are warmed rapidly in summer, and less rapidly in winter, and over both land and sea they absorb water vapour from the underlying surface (the rainfall of this region is sufficient to provide the water for evaporation from the land). In this process the air becomes increasingly unstable, and when lifted by the approach of a fresh incursion of cold air from the south (a cold front), it can give rise to showers or rain, often accompanied by thunder or hail.

Tropical air

The source region of tropical air masses is the subtropical high-pressure belt in latitude 20° to 35° South. The maritime tropical air masses of South America form in the areas of subsidence in the semi-permanent anticyclones of the Pacific and Atlantic oceans to the south of the trade-wind belt. Both these air masses have characteristics typical of a tropical maritime air mass near its source, i.e. a high water-vapour content and steep lapse rate in the lowest 2000–4000 ft, above which lies a marked inversion and a layer of warm, dry air.

Tropical continental air is formed by modification of polar air masses over the continent and has a much less stable character. Tropical maritime (Pacific) air forms the warm sector of the depressions which approach the coast of Chile from the west. It frequently reaches the coast of Chile causing coastal fog, low cloud or drizzle as it crosses the cold coastal current or as it is displaced southwards over colder water in advance of the Pacific depressions.

As the Pacific depressions usually occlude before crossing the Andes or passing Cape Horn, tropical maritime (Pacific) air is rarely found on the surface east of the Andes. A warm, very dry, high-level air stream is, however, frequently found crossing the Andes in latitude 30° to 35° South and may be regarded as the upper part of the tropical maritime air mass. It does not descend below 7000 or 8000 ft east of the Andes.

Tropical maritime (Atlantic) air flows west and south-westwards from the high-pressure centres off the east coast of Brazil. At a later point, moving south-eastwards it often forms the warm sector of the wave depressions which form over northern Argentina or Uruguay. As this air mass moves southwards the inversion of temperature, which was characteristic of the anticyclonic area in which it was formed, rises and decreases in intensity. Therefore, when the air is lifted in the circulation of the polar depressions it gives copious rainfall often of a showery character and accompanied by thunder.

Since the anticyclones in which this tropical maritime air mass is formed are frequently of a migratory type, moving north-east behind an outbreak of polar air, it is often difficult to decide when the air circulating round them should cease to be classified as returning polar maritime air and be regarded as tropical maritime air. The distinction is in fact one of degree, as the air of the moving anticyclone is gradually transformed from polar to tropical as the centre moves northwards and subsidence continues.

Tropical continental air
Tropical continental air is formed over the plains of northern Argentina, Paraguay and southern Brazil from the polar air masses, which frequently remain more or less stationary in these areas for several days. The process of modification is most rapid in spring and summer. It results in an air mass which is warm and moist in the lower layers, and relatively dry and cold aloft. Tropical continental air frequently moves south and enters into the circulation of the depressions, which form over Argentina. It produces showery precipitation with frequent thunderstorms.

Equatorial air

The origin of all equatorial air masses is in the trade winds of the northern and southern hemispheres. In the trade-wind belt, the air has the same characteristics as tropical maritime air, a moderately high water-vapour content and steep lapse rate up to a temperature inversion, below which the typical trade-wind cumulus forms. Above the inversion, the air is warm and dry as a result of subsidence in the tropical anticyclone. As the trade winds approach the Equator the temperature inversion becomes less pronounced, and the air becomes convectively unstable, giving rise to showers and thunderstorms when lifted in the inter-tropical convergence zone or heated over land. Equatorial air originating in the Atlantic trade winds of both the northern and southern hemispheres enters into the atmospheric circulation of South America.

In the southern winter (July), equatorial air from the North Atlantic affects only that part of South America north of the Equator, while equatorial air from the SE trades of the South Atlantic turns south-westwards and usually covers much of central Brazil. Its flow south-westwards is, however, irregular and although sometimes it may extend as far south as the extreme north of Argentina, at other times it may be replaced by air of polar origin almost to the Equator in central Brazil. It can also extend along the coast at least as far north as latitude 15° South.

In the southern summer (January), equatorial air from both hemispheres flows well southwards into Brazil, crossing the NE coast as roughly parallel north-easterly wind currents, separated by the ITCZ which oscillates from east to west across central Brazil.

It must be pointed out that the continental ITCZ is not so easily identified. Many texts indicate a neat unbroken line purporting to show where it is actually positioned. As the continental interior displays a vast low-pressure area, it is this that produces air-mass storm characteristics, rather than a more abrupt meeting of trade winds as seen over the oceans.

Equatorial air in this season covers most of Brazil, except the coastal region from $\cong 20°$ South. Rainfall statistics indicate a lighter rainfall in this region. This is probably due to the trade wind inversion in the south-east trades being still in evidence, and warm dryer air aloft which has not completely disappeared from this air mass in its passage over South America. This results in less instability showers and rain. Showers do not so readily develop, as is the case with equatorial air from the northern hemisphere, which has had a longer sea track from the source region in the North Atlantic trade winds.

Lower winds

Figures 26.6 and 26.7 show the stream lines of the lower winds $\cong 3000$ ft above mean sea level in January and July. Surface winds are very variable and there is no clearly defined predominating wind direction. Wind speeds are usually light or moderate. Sea breezes ('*Virazon*') develop regularly over the River Plate area, except during disturbed conditions, reaching their greatest development in the summer months. On the Uruguayan coast the sea breeze is an east/SE wind reaching 15 to 25 kt. In the Buenos Aires area it rarely exceeds 10 kt, and is generally between east and east/SE. The land breeze ('*Terral*') can also be observed on quiet mornings.

Gales are rare in the River Plate area but do occur near Montevideo with wind speeds of 33 kt or gusts of 42 kt. The winds occur either from west/SW ('*Pampero*') following a cold-front passage, or from east/SE ('*Sudestra*') usually in the circulation of a depression moving SE over Uruguay. Squalls of gale force that occur with the passage of cold fronts are quite frequent, but are short lived. Gales are almost unknown inland; squalls accompanying the passage of cold fronts occur particularly along the coastal areas.

Generally the whole South American Atlantic coastline from Venezuela to 30° South has predominantly NE to SE surface winds. To the south of 30° South in the disturbed temperate zone, strong winds are possible, variable in direction, in association with depressions and fronts ('*Roaring Forties*').

Surface winds west of the Andes are generally rather light and variable. In summer a westerly wind, the '*Travesia*', a combination of sea breeze and valley

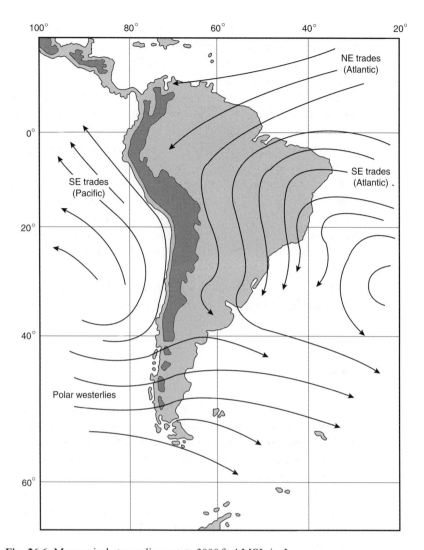

Fig. 26.6 Mean wind streamlines at ≅ 3000 ft AMSL in January.

breeze, develops with considerable regularity and blows inland in the late morning and afternoon. It reaches its greatest strength up the river valleys (≅ 25 kt at Juncal) but does not exceed 10 to 15 kt elsewhere. It is about 3000 ft deep and causes some turbulence over the coastal hills. Also in summer a fresh southerly wind, the '*Surada*', occasionally occurs when the thermal low over central Chile is well developed and pressure is high to the south.

Gales are rare, averaging three a year at Valparaiso. These are almost equally divided between exceptionally well-developed suradas in summer and northerly gales in advance of winter depressions. Gales inland are still less common.

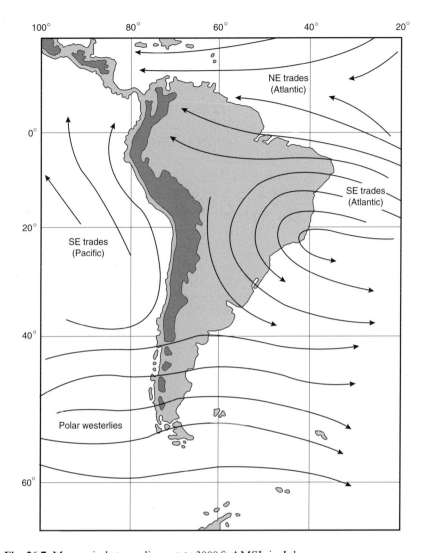

Fig. 26.7 Mean wind streamlines at ≅ 3000 ft AMSL in July.

26.2 General weather summary

Maximum rainfall west of the Andes occurs in winter (80 + % at Santiago). East of the Andes, the reverse is true, the dryness of the winter being most marked immediately east of the mountains. The summer and winter rainfalls are virtually equal on the east coast.

January weather

In the north and down to 15° South, the ITCZ will produce a great deal of thunder activity. Thunderstorms extend into southern Brazil with orographic cloud frequent along the coastline. In the south (Argentina) in March and April there occurs the highest rainfall from Atlantic polar front lows, but the sheltering effect of the Andes helps to reduce precipitation totals. Cloud amounts here are relatively small and marked Föhn effects and turbulence can be expected with strong westerlies.

During the southern summer, the northward passage of cold fronts give rise to heavy thunderstorms with squalls. This is due to the convectively unstable lapse rate in the warm air mass. The squalls which accompany these frontal rains and thunderstorms are known as '*Pampero sucio*' (not to be confused with the *pampero*, which is described in the July weather).

The squalls frequently approach from a south-westerly direction, have the character of a line squall and bring a large and sudden fall of temperature. They can be very violent, and it is estimated that in the Buenos Aires area they reach 60 kt once or twice a year. The squalls are associated with towering cumulus clouds and violent storms, producing torrential rain, sometimes hail, and also severe icing above the freezing level of $\cong 14\,000$ ft. Below the squall cloud visibility is frequently severely reduced by the very heavy rain. On some occasions, also, when the air is too dry for much rain to reach the ground, the squall raises dust from the dry and dusty surface, giving rise to a duststorms, but these are usually short lived; this is most common in spring over the western section of Argentina.

July weather

The most northern parts are influenced by the ITCZ. Heavy showers and thunderstorms are normal and produce a rainy season. South of the Equator to about 30° South, it is a dry season despite occasional isolated thunderstorms. From 30° South, it is true winter with much activity from the Atlantic polar front. Depressions and cold polar air can penetrate well into Brazil. This produces frontal weather with typical overcast skies, low cloud, and continuous rain in warm sectors with low stratus and drizzle.

A particular feature of northern Argentina is the *pampero*, a southerly (polar) wind which sets in with violent line squall characteristics following a cold front. It is similar in effect to the Australian 'southerly buster', and may carry clouds of dust from the pampas.

Before the arrival of a cold front over northern Argentina, winds are usually from a northerly direction. At the front they back sharply to SW, south, or even SE, and subsequently, as the following anticyclone or ridge of high pressure moves NE across the area, they swing round gradually to a northerly direction again. Whether the change is clockwise or anticlockwise depends on the path of the anticyclonic centre. The period occupied by the sequence varies considerably, the intervals between cold-front passages being from one to eight days.

The normal weather type (southerly type) in which cold fronts move NE across Argentina can thus be conveniently divided into three periods which are repeated with each front.

(1)   A period of northerly winds preceding its arrival.
(2)   The period of frontal passage.
(3)   The period of mainly southerly winds behind the front.

The Atacama (coastal) desert – note

This desert region lies to the north of Santiago $\cong$ 30° South to the Peruvian border ($\cong$ 18° South). This is one of the driest places on earth. It has not rained in this desert for 400 years (apart from a meteorological anomaly in 1971). The climate is dominated by the permanent high-pressure system over the Pacific. However, it is not cloud free. Medium and low cloud is often present, including mist, but rainfall is virtually totally absent.

Upper winds

January	10° North to 20° South	Easterly maximum 30 kt
	40° South to 50° South	Westerly maximum 50–60 kt
July	20° North to 10° South	Easterly maximum 30 kt
	40° South to 50° South	Westerly 55–65 kt
	40° South to 60° South	Jet stream 100 kt +

Freezing level

January	10° North to 20° South	16 000 ft
	40° South	10 000 ft
	50° South	< 8000 ft
July	10° North to 20° South	14 000 ft
	40° South	8000 ft
	50° South	< 5000 ft

Weather in the South Atlantic

The area under consideration is that part of the Atlantic which lies between the Equator and 60° South. See Fig. 26.8.

The semi-permanent anticyclone (part of the permanent high-pressure belt) is normally positioned between 20° and 30° South. Within that area, the winds are mainly light and the weather fine. The region of south-easterly trade winds lies on the northern edge of the high-pressure region, and south of 40° South are the 'Roaring Forties' – a belt of strong prevailing westerly winds which entrain depressions and frontal systems.

The climate of the South Atlantic is not unlike that of the North Atlantic but the presence of Antarctica and an unbroken southern ocean forcing a

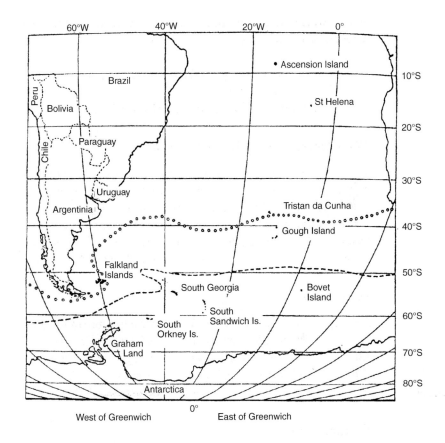

Fig. 26.8 South America and the South Atlantic. °°°°°°° Equatorial limit of icebergs; - - - - - - - Antarctic convergence zone.

strong sub-polar circulation, together with the lack of a feature comparable to the North Atlantic Drift, means that the climatic regions are displaced some 5° to 10° closer to the Equator. For example, the Falkland Islands (at about 51° to 52° South) have similar climatic features to the Shetland Islands (at about 60° North), being often rather cloudy and breezy but with good visibility. Seasonal variations are less pronounced over the South Atlantic than those experienced in the North Atlantic.

The Falkland Islands group lies in the region of strong westerly winds with fronts and depressions entrained. This region sees weather showing rapid deterioration and rapid improvements; the rugged, hilly terrain of the islands can cause dramatic local variations in weather. The Southern Andean Mountains also provides larger-scale influences from time to time. The unsettled flow is sometimes broken by a ridge of high pressure extending into the area but this feature rarely lasts as long as a week; consequently, there are few spells of settled weather.

The following weather elements are taken from the Stanley Data (*A Summary of the Climate of the South Atlantic*. Courtesy of the Royal Air Force, Finningley, October 1983.)

Pressure

Mean sea level (MSL) pressure monthly averages range from $\cong 1000$ mb in January (summer) to 1005 mb in September (spring). The day-to-day variation is greatest in the winter months. The extremes of MSL pressure at Stanley are 955 mb and 1039 mb.

Wind

The prevailing wind is from the west. At Stanley, winds are from 200°T to 340°T for 70 to 80% of the time. Winds blow only temporarily from other directions, usually in association with passing depressions. Wind speeds vary little throughout the year, averaging 14 to 17 kt in each month. Mean winds of 22 kt or more occur about one hour in four in most months of the year. Mean winds of 33 kt or more (gale force) occur on about 50 days per year, the highest hourly mean speed in an average year being 45 to 48 kt.

Mean wind speeds are likely to be higher over the open sea, exposed headlands and high ground, and extreme gusts may well be higher in the last two situations.

Mountain waves are a regular feature (on a large scale due to the Andes and on a smaller scale due to the island hills) and thus marked vertical currents may be encountered in smooth air. At low levels very severe turbulence is possible in the rotor zones downwind from high ground.

Precipitation

Stanley is thought to be one of the wetter parts of the Falkland Islands. Precipitation is frequent but not often in the form of heavy or prolonged rain showers. It is generally in the form of fairly light frontal rainfall or light to moderate rain showers. The average annual rainfall in Stanley is about 610 mm, i.e. similar to that for the London area.

Snow or sleet falls in every month of the year. During the winter (June to August) it can be expected to occur on 10–11 days per month. Significant accumulations are rare, however, the greatest observed depth being $\cong 15$ cm on low ground.

Hail can be expected on one to three days per month throughout the year but thunder is almost exclusive to the summer, occurring on one to two days per month from December to February.

Temperature

The mean annual temperature at Stanley is 6°C, the warmest month being January and February at $\cong 10$°C and the coldest July at 2°C. Temperature ranges are restricted by persistent cloud and wind but a temperature of 25°C has been recorded in January. Spells of severe cold are rare, even in mid-winter, temperatures below $-7°C$ being uncommon.

Frequent precipitation and moderate to fresh winds together can produce a much lower 'effective' temperature and on high ground with increased wind speeds and lower temperatures, conditions can be very severe. Daytime relative humidity ranges from 80 to 90% in winter and 70 to 80% in summer.

Visibility

On low ground and away from windward coasts, visibility exceeds 10 km for about 90% of the time in summer and 80% of the time in winter.

Sea fog is common around the Falkland Islands but inland fog is rare, the visibility there being more often impaired by precipitation. Hill fog is not uncommon, however, and poor visibility is thus more likely with increasing altitude.

Cloud and sunshine

Mean cloud cover at Stanley is 5/8 to 6/8 throughout the year, the winter being slightly cloudier than summer and daytime being slightly cloudier than night. Over high ground, orographic cloud often occurs, especially on windward hillsides, and thus low cloud can be expected to obscure high ground fairly frequently. Mountain wave clouds are often observed, occasionally stacked above each other like plates.

The Falkland Islands are surprisingly sunny, the annual average being 1618 h, which compares favourably with the south coast of England. December has the highest average sunshine with 218 h and June the lowest with 55 h.

Icing risks (based on data from Stanley)

Winter Aircraft icing is a fairly common hazard over the far south of the South Atlantic. There is a risk of icing in the lowest 4000 ft during at least part of 20 days per month. On one day in three, the risk in shower cloud would be moderate or severe but limited in extent. On other days the icing may be extensive.

Summer The risk of icing in the first few hundred feet is very low but still possible on one or two days per month. At about 3000 ft, the probability of icing is about 25% of the winter values.

Weather on Ascension Island

Ascension Island is a small and mountainous island positioned in mid-Atlantic at about 8° South, 14°15′West (see Fig. 26.8). The island lies within the SE trade wind belt and seasonal climatic variations are small. The ITCZ does not extend as far south as Ascension Island.

The following climatalogical data has been taken from three sources:

Georgetown (62 ft AMSL), Wideawake Field (272 ft AMSL, about 2.5 km SE of Georgetown), and Green Mountain (2818 ft AMSL situated on the east side of the island).

Wind
At Georgetown winds are easterly or south-easterly for over 80% of the time, and on 90% of occasions the speed is 10 kt or less. Strong winds of over 20 kt are infrequent and gales do not occur. At the eastern (windward) end of the island, however, winds are likely to be fresh or strong more frequently and gales, although uncommon, have been observed.

Precipitation
Average annual rainfall ranges from 142 mm at Georgetown and 173 mm at Wideawake to 657 mm on Green Mountain. Exceptional rainstorms affect Ascension Island (usually during March or April) for long intervals. They may yield 150 to 200 mm, about a year's average rainfall in 24 hours or less. Snow is unknown and thunder or hail is unlikely at any season.

Temperature and humidity
The warmest months of the year in Georgetown are April and May, averaging 32° to 33°C by day and 24°C by night. The coolest months are August to October, about 29°C by day and 21°C by night. The temperature excursion is thus small.

Relative humidity
This varies little during the day or seasonally, averaging 67 to 75% by day and 75 to 80% by night.

Visibility
Visibility is generally good on Ascension Island, ranging from 1 to 4 km around dawn to 10 km or more by noon on most days. The visibility rarely falls below 1000 m at any time on low ground but Green Mountain is affected by hill fog from time to time although this hill fog rarely extends below 1000 ft.

Cloud and sunshine
Completely clear skies or overcast conditions are rare but Green Mountain is often cloud-capped. The cloudiest months at Georgetown are September to November with 5 to 6 oktas of cloud in the morning decreasing to 4 to 5 oktas by evening. February and March are the least cloudy months with 2 to 3 oktas throughout the day. Throughout the year the cloud base is usually between 2000 and 5000 ft. Flying conditions are generally very good for most of the year.

Chapter 27

Route Weather Dakar to Recife

27.1 The Equatorial climatic zone route

The characteristic is two main wet seasons associated with the passage of the sun north and south across the Equator, but there is no real dry season. There will be a great deal of cloud along the convergence zone (ITCZ) with cumulus and heavy showers and frequent thunderstorms. Temperature and humidity is high and almost constant thoughout the year.

The ITCZ never moves further south than about 2° South. The trade winds cover this part of the Atlantic Ocean, between the subtropical high-pressure belts and the Equatorial trough. Therefore, north of the Equator they generally blow from the NE and south of the Equator they blow from the SE. When the ITCZ is located relatively far from the Equator, the portion towards the Equator is occupied by the equatorial westerlies, i.e. the deflected trade winds of the opposite hemisphere.

The trade winds become unstable in the lower levels while flowing equa-torwards over the sea. The presence of the trade wind inversion, however, restricts the tops of the trade Cu to about 10 000 ft. The consistency of the trade winds is remarkable throughout the year.

The upper winds are westerly 10 to 25 kt over the whole route in January, but are easterly about 10–15 kt over most of the route in July, becoming light westerly near Dakar.

The weather at Recife is associated with the trade winds which blow from the SE and bring in considerable rainfall, but the mountains behind the coastal region rise to about 3000 ft and effectively stop the surface moist layer from penetrating inland. During the southern winter, Recife receives most of its rainfall from weak troughs in the trade-wind circulation. Cloud bases less than 1000 ft and visibility less than 5 km are almost always associated with southerly winds.

The weather at Dakar is dependent on the position of the ITCZ. In January, the ITCZ is furthest south and the trade wind inversion is about 4000 to 5000 ft in depth, which restricts any great development of Cu. The NE wind blowing from the Sahara (the Harmattan) is dust laden, and is a hot dry wind; the air is hazy – visibility is best just before dawn, but can deteriorate after sunrise.

During July/August, the weather at Dakar is influenced by the fact that the ITCZ has moved to the north, and therefore the area is under the influence of the south-west monsoon; the rainfall is greatest during this period, and the layer of moist air is deepest allowing thunderstorms to develop. There is a great deal of medium cloud before dawn, and low stratus usually forms at sunrise and soon after. Later, the medium cloud breaks up, but by this time, Cu and Cb are developing inland.

This is the season of the West African line squalls, although these are mainly restricted to the coastal regions of Ghana and the Ivory Coast, but they can extend to affect the Guinea coast. They move towards the west and out into the Atlantic, and can be a hazard on the route from Recife. The extent of these line squalls does not always affect Dakar, but when it does, a distinct squall line is observed, and presents a hazard to pilots. There are two main squall seasons, one as the ITCZ moves north and another as the ITCZ moves south.

- Tropopause height is virtually constant at about 56 000 ft throughout the year.
- Freezing level is the same as for the general tropical areas, 16 000 ft.
- There are no tropical cyclones on this route, as they do not form south of the Equator in the Atlantic, and to the north of the Equator on this route the sea is too cold.
- The mean wind component on the route Dakar to Recife is a tail wind of 14 kt at 30 000 ft and 16 kt at 40 000 ft.
- See Fig. 5.4 for a satellite image of the ITCZ across this route.

Chapter 28

El Niño and La Niña

In this chapter, extracts are taken from a paper by D Painter (1997) and papers by the NOAA. The former is incorporated with the kind permission of the author and the Remote Imaging Group. Acknowledgement is also made of the contributions by the NOAA.

'El Niño' is Spanish for 'the little boy or Christ child'. The term refers to the appearance in early December of a dramatic change in the climate of the Pacific coastline and waters off Peru, the effects of which herald the arrival of the El Niño. The events that unfold following the arrival can propagate around the globe causing dramatic changes in local and world weather conditions.

28.1 The cause

El Niño is characterised by a large-scale weakening of the trade winds and warming of the surface layers in the eastern and central Equatorial Pacific Ocean. El Niño events occur irregularly at intervals of 2–7 years, although the average is about once every 3–4 years. They typically last 12 to 18 months, and are accompanied by swings in the southern oscillation (SO), an interannual seesaw in tropical sea-level pressure between the eastern and western hemispheres. The El Niño in some years, notably in 1983, caused the reversal of the eastern trade winds in the southern hemisphere!

28.2 The basic theory

The world's oceans cover approximately 70% of the planet, and the Pacific Ocean is the largest ocean, and as a result it is responsible for controlling many aspects of the global climate. Its deep-water currents are usually stable and predictable, but in El Niño years the usually warm western Pacific waters flow farther south and east than normal, and this puts a large relatively shallow mass of warm water on top of the normally colder southern Pacific waters.

During El Niño, unusually high atmospheric sea-level pressures develop in the western tropical Pacific and Indian Ocean regions, and unusually low sea-level pressures develop in the south-eastern tropical Pacific. SO tendencies for unusually low pressures west of the date-line and high pressures east of the date-line have also been linked to periods of anomalously *cold equatorial Pacific sea surface temperatures (SSTs)* sometimes referred to as *La Niña*. The southern oscillation index (SOI), defined as the normalised difference in surface pressure between Tahiti, French Polynesia, and Darwin, Australia, is a measure of the strength of the trade winds, which have a component of flow from regions of high to low pressure.

High SOI (large pressure difference) is associated with stronger than normal trade winds and La Niña conditions, and low SOI (smaller pressure difference) is associated with weaker than normal trade winds and El Niño conditions. The terms 'El Niño southern oscillation event' (ENSO) and 'ENSO cycle' are used to describe the full range of variability observed in the southern oscillation index, including both El Niño and La Niña events.

The trade winds transport vast volumes of warm surface sea water from the eastern to the western Pacific Ocean. The strong easterly trade winds in 'normal' or non El Niño years effectively pile up or tilt the sea surface ever so slightly up at the western end of the Pacific causing a gradient. The cold sea water from deep below wells up to replace the 'transported' surface water and produces an 'upwelling' of cold sea water off the Peruvian coast.

In El Niño years, the easterly trade winds decline, the sea gradient levels off and the warmer surface water very slowly moves back across the Pacific to the Peruvian coast. This creates the warm water cap over the coastal upwelling and produces the El Niño.

The change in the trade winds causes complex interactions within the sea's differing thermoclines, tidal currents and upwelling. All these areas play a part in the sea's ability to absorb heat from the sun, and the changes in sea current patterns and temperature propagate unusual subtle effects around the globe causing some of the global weather changes. The shifting of the patterns of rainfall causes the flooding and prolonged droughts that affect every part of the globe including the polar ice-caps.

The TOPEX/Poseidon satellite can observe long-term changes in sea level, sea surface temperature (SST) and in the temperature gradients or thermoclines across the world's oceans, particularly the Pacific Ocean. These gradients or thermoclines are layers within the sea that have differing salinity and temperature. They also 'slope' from hotter, shallow water into deeper, colder water (and vice versa).

El Niño's 'companion' La Niña *is a warming* of the Western Pacific Ocean, *while the Eastern Pacific gets cooler*. La Niña also effects the weather but to a lesser degree than El Niño and it does not always follow that a La Niña will follow an El Niño.

The effects of La Niña tends to be opposite on the global climate to those of El Niño. In the tropics, ocean temperature variations in La Niña tend to be opposite to those of El Niño. At higher latitudes, El Niño and La Niña are among a number of factors that influence climate. However, the impacts of El Niño and La Niña at these latitudes are *most clearly seen in wintertime*. In the continental United States, during El Niño years, temperatures in the winter are warmer than normal in the north central states, and cooler than normal in the south-east and the south-west. During a La Niña year, winter temperatures are warmer than normal in the south-east and cooler than normal in the north-west.

28.3 Predicting the event

The NOAA and the United States National Weather Service have a long chain of weather buoys across the Equatorial Pacific called the 'TAO ARRAY' (tropical atmospheric ocean array) gathering 'real time' information on the progress of the El Niño, as well as the TOPEX/Poseidon, NOAA, GOES and SeaWiFS satellite programmes. The NOAA is also closely involved collating data and providing it to interested parties and governments, as it has a particular interest in predicting the hurricanes and tornadoes off the United States coast.

Many countries watch the signs of an impending El Niño, so much so that watching 'the El Niño' is now a major full time occupation of many scientific institutes and governments, as its effects can devastate whole economies. The United States is one of the countries that are using a wide collection of techniques to help predict and track the progress of El Niño.

There is some very strong evidence to support the fact that in the El Niño years there are some of the worst-ever recorded weather events in the United States and the Caribbean. As the NOAA provides the early warning services for the United States and several other countries about the development of storms and tropical storms, it takes a keen interest in following the latest data.

The 1997 El Niño

The effects of the 1997 El Niño have proved that this had the makings of the worst-ever recorded El Niño, even worse than the 1982/1983 event. Comparisons of previous events have shown that the quicker the changes following the appearance of the unseasonable warm waters off Peru, the worse the succeeding weather.

Reference

Painter, D. (1997) El Niño – The unwanted Christmas gift. *Journal of the Remote Imaging Group* **51**, 31–36.

Appendix 1

The Beaufort Wind Scale

Land scale

Beaufort force	Descriptive term	Speed equivalent		Specification
		Mean kt	Gusts kt	
0	calm	< 1		Calm; smoke rises vertically
1	light air	1–3		Direction of wind shown by smoke drift, but not by wind vanes
2	light breeze	4–6		Wind felt in face; leaves rustle; ordinary vanes moved by wind
3	gentle breeze	7–10		Leaves and small twigs in constant motion; wind extends a light flag
4	moderate breeze	11–16		Raises dust and loose paper; small branches are moved
5	fresh breeze	17–21		Small trees in leaf begin to sway; crested wavelets form on inland waters
6	strong breeze	22–27		Large branches in motion; whistling heard in telegraph wires; umbrellas used with difficulty
7	near gale	28–33		Whole trees in motion; inconvenience felt when walking against the wind
8	gale	34–40	43–51	Breaks twigs off trees; generally impedes progress
9	strong gale or severe gale	41–47	52–60	Slight structural damage occurs; chimney pots and slates removed
10	storm	48–55	61–68	Seldom experienced inland; trees uprooted; considerable structural damage occurs

Beaufort force	Descriptive term	Speed equivalent		Specification
		Mean kt	Gusts kt	
11	violent storm	56–63	69–77	Very rarely experienced; accompanied by widespread damage
12	hurricane force	> 64	> 78	–

Notes
(1) Beaufort force 9 is described as 'severe gale' in the United Kingdom in preference to the WMO description of 'strong gale'.
(2) Gust criteria are used by the United Kingdom Meteorological Office but are not included in the WMO scale.

Sea scale

Beaufort force	Descriptive term	Speed equivalent	Specification
		Mean kt	
0	calm	< 1	Sea like a mirror
1	light air	1–3	Ripples with the appearance of scales are formed, but without foam crests
2	light breeze	4–6	Small wavelets, still short but more pronounced. Crests have a glassy appearance, but do not break
3	gentle breeze	7–10	Large wavelets. Crests begin to break. Foam of glassy appearance. Perhaps scattered 'white horses'
4	moderate breeze	11–16	Small waves, becoming longer; fairly frequent 'white horses'
5	fresh breeze	17–21	Moderate waves, taking a more pronounced long form; many 'white horses' are formed, chance of some spray
6	strong breeze	22–27	Large waves begin to form; the white foam crests are more extensive everywhere. Probably some spray

(Continued on p. 254)

(Continued.)

Beaufort force	Descriptive term	Speed equivalent	Specification
		Mean kt	
7	near gale	28–33	Sea heaps up and white foam from breaking waves begins to be blown in streaks along the direction of the wind
8	gale	34–40	Moderately high waves of greater length; edges of crests begin to break into spindrift. The foam is blown in well-marked streaks along the direction of the wind
9	strong gale or severe gale	41–47	High waves. Dense streaks of foam along the direction of the wind. Crests of waves begin to topple, tumble and roll over. Spray may affect visibility
10	storm	48–55	Very high waves with long overhanging crests. The resulting foam, in great patches, is blown in dense white streaks along the direction of the wind. On the whole, the surface of the sea takes on a white appearance. The 'tumbling' of the sea becomes heavy and shock-like. Visibility affected
11	violent storm	56–63	Exceptionally high waves (small and medium-sized ships might be for a time lost to view behind the waves). The sea is completely covered with long white patches of foam lying along the direction of the wind. Everywhere the edges of the wave crests are blown into froth. Visibility affected
12	hurricane force	> 64	The air is filled with foam and spray. Sea completely white with driving spray; visibility very seriously affected

Appendix 2

Summary of Local Winds

Anabatic wind   The term 'anabatic' is applied to local winds. The conditions to produce such winds are the reverse of those in the case of katabatic winds. During a clear day the upper surfaces of mountains, due to the greater transparency of the atmosphere at altitudes, are subjected to a higher degree of insolation than are the surfaces of the lowlands. The air in contact with the upper mountain surfaces is, therefore, warmer than that in contact with the surfaces of the lowlands and, as a result, convectional currents pass up the slope of the mountains. This also assumes that the air is intrinsically stable.

Baguio   A local name by which the *tropical cyclones* in the neighbourhood of the Philippines are known, particularly those that affect the area from July to November.

Barat   A squally, occasionally violent and damaging wind blowing across the Celebes Sea to the north-east coast of the island of Celebes.

Barber   A term used in some sections of the United States and Canada to describe a strong wind which carries precipitation that freezes upon contact with objects, especially the beard and hair.

Belot or belat   A strong land-wind from the north and north-west along the south-east areas of Saudi Arabia from December to March. Dust picked up by this wind often produces a hazy appearance to the sky.

Blizzard   The UK Met Office defines a blizzard as the simultaneous occurrence of moderate or heavy snowfall with winds of at least Beaufort scale 7 (28– 33 kt), causing drifting snow and visibility to 200 m or less. A severe blizzard implies winds of at least Beaufort scale 9 (41–47 kt, gusts 52–60 kt) and a reduction of visibility to virtually zero.

Bora   A cold and often dry north-easterly wind down the mountain slopes of the eastern coasts of the Adriatic, particularly in the north. It often produces violent gusts. It is mainly a winter phenomenom, but does occur in summer.

Brickfielder or brickfelder   This is a hot, dry and dusty wind experienced in Australia. It blows over southern Australia from the central desert regions. The name originated in Sydney because of the dust raised from the brick fields south of the city.

Bull's-eye squall   A sudden squall forming in apparently fair weather, characteristic over the ocean off the Cape of Good Hope, South Africa. Its name is derived from the appearance of a small cloud marking the top of an otherwise invisible vortex of the storm.

Burga   A strong windstorm in Alaska, usually attended by snow or sleet.

Cacimbo   A refreshing, cooling sea breeze blowing almost daily in July and August from the south-west to the port of Lobito on the coast of Angola in western Africa.

Chergui   An intrusion of hot air (easterly wind) into Morocco. Originates in the Sahara Desert.

Chubasco   A violent squall-type wind associated with severe thunderstorms, frequently occurring on the western coastal sections of Central America and Mexico.

Cockeyed bob   A squall wind associated with thunderstorms on the north-west coast of Australia, blowing most frequently from December to March.

Contrastes   Winds that blow from opposite directions even though a short distance apart. The western Mediterranean area is frequently subjected to these contrasting winds in the spring and autumn seasons.

Coronazo   Strong south winds blowing from the west coast of Mexico. These winds are usually the eastern peripheries of tropical storms located well offshore to the west.

Depeq   Local name to the strong winds over Loet Tawar (Sumatra) during the south-west monsoon.

Dimmerföhn   A rare form of Föhn wind (see also entry for Föhn wind) where during very strong upper winds from the south a pressure difference of the order of 12 mb (or more) exists between the south and north of the Alps (Europe). A stormy föhn wind then overlaps the upper valleys in the northern slopes, reaches the ground in the lower parts of the valleys, and enters the foreground as a very strong wind. The Föhn wall and the precipitation area extend beyond the crest across the almost calm surface in the upper valleys.

Doctor   A term originating in England to describe cooling sea breezes occurring in the tropics. More a generic name, as 'the doctor' can be found in several countries.

Dust devils (or dust whirl or dust pillar)   The whirlwinds of dry, sandy regions are often referred to as dust devils. These convection currents carry dust into the air and the direction of rotation may be clockwise or anti-clockwise. They often occur in desert or semi-desert regions, and vary in velocity from about 3 kt to over 50 kt. They usually reach about 30 m in height, but have been known to reach as high as 1000 m.

Elephanta   A strong wind blowing from the south or south-west on the extreme south-west end of India during September and October. It marks the beginning of the dry season and the end of the rainy south-west monsoon.

Etesian (Meltemi)   This is a katabatic north-easterly wind, which is prevalent over the Grecian archipelago during summer, particularly over the Aegean Sea, where it becomes more northerly.

Föhn wind (or Foehn wind)   A wind characteristic of many mountainous regions of the world and called by different names in different countries (Chinook, Santa Ana, etc.). The Föhn wind is a warm, dry wind, which blows down the slopes on the leeward side of a ridge of mountains. The name originated in the Austrian Alps where the Föhn is very prevalent, especially on the northern slopes. The wind on reaching the windward side of the mountains ascends, and the dynamic cooling which results causes cloud formation (of lenticular shape) and precipitation. The air then passes down the leeward side, and having lost its moisture and being heated by adiabatic compression it reaches the valley below as a warm dry wind. Föhn winds are likely to occur wherever cyclonic systems pass over mountain regions. As an example, they occur frequently on the coast of Greenland where they exert considerable influence on the winter climate.

Gharbi   Warm, moist southerly winds affecting the Adriatic and Aegean Sea regions. Dust is carried all the way from the Sahara and is often mixed with rain causing 'red rain'.

Gregale   The gregale is a strong north-easterly wind which blows over the Ionian and Mediterranean Seas, especially during winter. It usually lasts two or three days and reaches gale force.

Haboob   This is the name given to a dust storm in the Sudan, May to September, especially near Khartoum.

Harmattan   This is a very dry wind common in Western Africa during the dry season (November to March). During these months, the winter of the northern hemisphere, the air over the Sahara cools rapidly because of its clearness and lack of moisture, and as a result becomes denser. It then flows outwards to the coast, especially to the Gulf of Guinea, to replace the warmer and lighter air. Since it is both dry and comparatively cool, it forms a welcome relief from the continuous damp heat of the tropics. It has also been named as the 'doctor' in spite of the fact that it carries with it from the desert large quantities of fine dust which penetrates everywhere, and can reduce the visibility to 1000 m or less. The visibility is often better in the early mornings.

Helm wind   Strong north-east winds in the Pennine chain in north-central England. These winds are sometimes associated with a roll or series of rolls of clouds that overhang the crest of a wind-wave to leeward of the hills. The mountain-top clouds that form in the wind are called helm clouds.

Imhat   A refreshing sea breeze that tempers the heat of the North African coastal areas.

Karaburan   From early spring to late summer, these gale-force winds form daily in the Gobi Desert and surrounding regions. Blowing from the ENE, they carry clouds of dust from the desert. The blowing dust often darkens the sky and it the reason the karaburan is sometimes called the black storm.

Katabatic wind   From high ground there is more radiation than from valleys, as there are fewer dust particles and water vapour particles to intercept it. As a result, the land becomes colder and the air in contact with the high land also becomes colder. The colder air is more dense than the surrounding air, and due to gravity flows down the slope. These winds often have no relation to the distribution of atmospheric pressure, and are purely local in character. Usually they are quite gentle, but if the valley path is long and steep and the summit is kept cold by snow, gale force winds results.

   A good example of a katabatic wind is the *Bora*. This cold north-east wind blows down from the plateau to the north of the Adriatic. The plateau becomes very cold in clear weather in winter, and a current of air descends into the warmer valley. When the Bora is reinforced by the northerly winds in the rear of a depression which are travelling eastwards along the Mediterranean, the winds blow with speeds up to 70 kt.

Another katabatic wind, less violent than the Bora, is the *Mistral*, which flows along the Mediterranean coast of France as far as Genoa, but is most prevalent between Montpellier and Toulon. The Mistral results from the combination of the snow-covered plateau of south-eastern France and the low barometric pressure in winter over the Gulf of Lyons. It is cold, dry, northerly or north-westerly, and is usually accompanied by clear weather and bright sunshine. Powerful katabatic winds also blow from the snow and ice-covered plateaux in central Greenland.

Khamsin   This is a south-easterly to south-westerly wind blowing over Egypt in front of depressions and passing eastward along the Mediterranean. This wind is hot and dry and produces duststorms, and sandstorms if the wind is greater than 16 kt. It is frequent from April to June. The name 'khamsin' is also given to the gales (dust- or sand-laden) in the area of the south or south-west of the Red Sea. Most prevalent during July to September.

Kharif   A strong, often gale force wind blowing from the south-west in the Gulf of Yemen. It is called the kharif on the Somaliland coast on the south shore of the Gulf, where the wind descends, dust- or sand-laden, and hot from the African interior.

Koshava   A stormy north-east wind in Yugoslavia, carrying snow from Russia.

Levanter   This is a hot, damp, easterly wind, which blows over Gibraltar during the summer. It occurs as a north-easterly wind on the east coast of Spain, especially from February to May and from October to December. In winter, it is strong and squally and may continue for two or three days. In summer, it is weaker and does not last so long. It results from high pressure over Central Europe and a depression over the south-western Mediterranean or African coast.

Marin   Southerly winds in the Gulf of Lyons blowing ahead of a depression. Very moist, therefore produces low stratus and drizzle. Very warm and humid.

Mistral   The 'masterful', a northerly wind of the Gulf of Lyons. Surges of polar air across north and central Europe affect a wide area of the north-west coast of the Mediterranean. It is a cold and dry wind and becomes very strong, 70 kt has been recorded. It is particularly noted in winter in the lower Rhône Valley.

Monsoon   (A season). Usually applied to the winds in the tropics. The term is commonly used in Asia and the Arabian Sea. The south-west monsoon is a rainy season, and the north-east monsoon is dry. The term has become generic to mean a very wet spell.

Narai   A cold wind in Japan, from the north-east and polar regions of the Asiatic landmass.

Northeaster   A wind that blows from the north-east over the New England and middle Atlantic coastal regions of the USA. This wind from the North Atlantic is generally moist and often chilly or cold. Cloudiness and precipitation frequently accompany it.

Northwester   A moderate to strong wind from the north-west bringing cool to cold temperatures over broad regions east of the Rockies. The name is also applied to frequent gale winds that batter the Cape region of South Africa.

Norther (or Norte)   A cold and strong northerly wind which blows over the eastern coast of the Gulf of Mexico during the months September to March. Along the coast it is sometimes humid and precipitation takes place. When the northers cross southern Mexico and the Sierra Madre into the Gulf of Tehuantepec, the wind is cold and dry, and can set in suddenly.

Oe   A localised whirlwind off the coast of the Faroe Islands in the north-east Atlantic.

Pampero   The pampero is a violent south-westerly line squall over the Rio de la Plata (Argentina/Uruguay) during the period July–September. It is sometimes accompanied by thunder and lightning. It is attended by cold fronts as they move from the south-west to the north-east over the pampas. It resembles the norther of the United States plains area in that it is an outbreak of air from polar latitudes. There is a marked drop in temperature when it passes.

Ponente   A westerly wind over the Mediterranean, particularly as a refreshing sea breeze on the western Italian coastline.

Purga   Another name for the buran of the tundra regions in northern Siberia. The purga sweeps down from the north with considerable violence.

Ravine wind   A wind which blows through a ravine or valley penetrating a mountain barrier. This would be the easiest route. There will be a pressure gradient either side of the mountain barrier; these winds can assume great strength because of the funnelling action. An example of a ravine wind is that which occurs at Genoa, and is caused by the pressure gradient between the Po valley and the Gulf of Genoa.

Reshabar   A strong wind blowing from the north-west over the Caucasus Mountain range between the Black and Caspian Seas.

Roaring forties (and Howling fifties)   A term probably originating with whalers and others of the nineteenth century who sailed the oceans of the southern hemisphere. They found the winds to be very strong and very consistent on their routes southward. The winds circle the globe at latitudes 40° to 50° South, with very little landmass to impede progress. The winds are the prevailing westerly belt that is the Earth's primary atmospheric circulation.

Seistan   The seistan is a northerly (downslope) wind, which is prevalent in Seistan (Afghanistan) during the summer (May to September). It has been known to reach hurricane force, and is often referred to as the wind of 120 days, owing to the fact that it continues for this period of time.

Shamal   This is a general north-westerly wind, hot and dry, which blows over Iraq and the Arabian Gulf at intervals during the period December–April. It displays marked diurnal variation.

Siffanto   A warm southerly wind blowing from the 'heel' of Italy.

Simoon or Simoo   A sirocco type wind, hot, dry, and dust-laden, blowing over the middle and southern portions of the Mediterranean. It is called a simoon when it is abnormally strong over the south-eastern part of the Mediterranean. The wind blows northwards originating in the hot desert sections of north-central Africa. The Turkish version of this wind is the *samiel*. The name simoon is also given to a hot whirlwind which frequently occurs in the African and Arabian deserts during spring and summer. If often carries large quantities of dust and sand, if the wind is strong enough, and lasts for about 20 minutes.

Sirocco (or Scirocco)   This is a southerly or south-westerly wind blowing in front of an advancing depression travelling west to east along the Mediterranean. The sirocco passes over the Sahara Desert, and therefore arrives at the north coast of Africa as a hot dry wind. Its temperature is further raised in its descent from the inland plateau to the coast. In its passage over the Mediterranean it takes up water vapour, and reaches Malta, Sicily, Italy and other parts of the European coast as a warm, moist wind. Local names for this wind are the *chilli*, blowing from the Algerian/Tunisian coast, and the *ghibli* blowing from the Libyan coast. However, the sirocco is the general name for all winds blowing from North Africa.

Sno   Cold, swift-moving currents of air from the highlands that fill the Scandinavian valleys in the winter attaining considerably strong velocities in the fjords.

Southeaster   Winds from the south-east, sometimes of gale force, that blow near the extreme south-west end of the Cape of Good Hope, South Africa. They usually occur in the winter and often bring a whitish haze of salt particles and sea spray.

Southerly burster   The southerly burster blows over New South Wales (Australia) and is associated with depressions. The air moving in from behind cold fronts is accompanied by strong to gale-force winds. The cold winds originate from the polar zones and move northwards. The gales are often orographically intensified over the highlands of New South Wales on the south-east coast; a great drop in temperature results. Storms may last between a few hours and several days, and are often accompanied by thunder and lightning.

Steppenwind   A cold north-east wind that occasionally blows over Germany from the steppe regions of Russia.

Sudeslades or Suestado   Strong to gale-force winds from the south-east that affect coastal areas of Uruguay, Argentina and Brazil, accompanied by cloudiness and rain.

Suhaili   A strong wind from the south-west blowing over the Arabian Gulf, bringing thick clouds and rain.

Sumatra   The sumatra is a south-westerly squall which is experienced in the Straits of Malacca, usually during the period April–October (south-west monsoon). It is usually accompanied by thunder, lightning and heavy rain, and the cloud is in the form of a bank of cumulonimbus.

Surazos   Strong cold polar winds of the Andes Plateau in Peru, with intensifying velocities as they sweep through the mountain passes. The surazos usually bring temperatures below freezing with clear skies.

Taku   A strong wind from the east or north-east blowing in the vicinity of Juneau,

Alaska. The name is taken from the Taku River. At the mouth of the Taku the wind sometimes reaches near hurricane force.

Tehuantepecer A violent cold wind from the north affecting the region around the Gulf of Tehuantepec on the extreme south coast of Mexico. The wind is intensified by a mountain valley funnel effect and sweeps southwards over the Gulf at gale strength. The onset has few if any precursory indications.

Trade winds Sometimes described as tropical easterlies these are winds which diverge from the subtropical high-pressure zones, centred at $\cong$ 30° to 40° North and South. They blow towards the Equator, are north-easterly in the northern hemisphere, and south-easterly in the southern hemisphere.

Tramontana A northerly wind in the Mediterranean. A local name. The wind is usually dry and cold.

Vardarac (or Vardar) A similar wind to the bora. Northerly wind blowing through the Morava–Vardar gap in the rear of a depression. Affects Thessaloniki in Greece in the winter. The wind is very strong and squally. It is a type of ravine wind.

Vendevale A strong south-west to westerly wind in the Straits of Gibraltar. It blows ahead of a cold front crossing Spain, mostly September to March, and is very squally with a great deal of low cloud.

Viuga Stormy north-east winds of southern Siberia, associated with passing low-pressure areas.

Whirly A small but violent storm in the Antarctic. The whirling winds of the storm may cover an area of up to 100 m or so in diameter, occurring most frequently near the time of the equinoxes.

Zephyr A soft, gentle breeze, mostly from the west of the Mediterranean region. (More mythological than real!)

Weather at Selected Destinations

Athens   In summer the *Etesian* may bring strong winds from a northerly direction. They tend not to be very steady in direction and make approaches and landing difficult. This is particularly the case in the summer. Strong squally winds from the east cause similar problems. The visibility is more often good at the airport. Apart from strong squally easterly winds there are no other serious weather problems.

Auckland (New Zealand)   A disturbed temperate climate. Weather controlled by a succession of eastward moving anticyclones and depressions. Cyclones in the West Pacific may reach New Zealand between December and March. With a moist north-easterly wind, the cloud base may lower to 300 ft, and visibility reduce to < 1000 m. General clearance occurs after the frontal passage if the wind backs to the south-west.

Azores (Larges)   The airport lies in the permanent high-pressure belt, or is influenced by the NE trades; however, only during the summer can good weather be expected. During winter and spring, the Azores are affected by the passage of depressions and generally disturbed weather. This produces varied weather and stormy and wet conditions. A further hazard is the wind effects at the airport due to topography.

Bahrain   Radiation fog occurs in autumn and winter but is seldom persistent. In winter, lows from the Mediterranean may bring some rain. From June to August visibility is often reduced by dust and for long periods there may be considerable haze.

Beirut   The airport receives shelter from easterly winds by the high ground in the east. However, there can be a strong katabatic wind of up to 20 kt flowing from the NE on summer nights. Thundery squalls can bring in rain and surface winds gusting to 30 kt and this can raise sand reducing visibility to less than 50 m. Fog is unknown.

Bermuda (Kinley Field)   The airport has a good weather record. Heavy rain produces the lowest visibility. During winter there are gales, and these can be severe. During the summer, tropical storms may pass through the area; this is mainly during August to November. They can affect Bermuda if close enough to give winds of gale force, but full hurricane winds are rare.

Bombay (Santa Cruz)   Fog at night during the NE monsoon in February and March. Frequent thunderstorms or heavy showers during the period of the south-west monsoon in June, September, October or to a lesser extent in May. There is a marked sea breeze from the west. The land breeze is quite marked allowing landings towards the east, but not during the monsoon.

Brazil (Recife)   The airport is subject to a great deal of cloud and rain during the wet season (March to August). In addition, low cloud base and poor visibility always occur with winds from a southerly direction.

Buenos Aires   Rain mainly during the summer months. Mild conditions during winter, moderated by the warm Brazil current. In January, the mean daily temperature is 25°C and average precipitation 76 mm. Thunder is heard about 7 days a month. The MSL pressure is 1012 mb. In July, the mean daily temperature is 10°C, average monthly precipitation is 53 mm. Thunder is heard about 3 days a month. MSL pressure is 1019 mb. Surface winds are very variable, and there is no clearly defined dominant wind direction. Sea breezes (virazon) develop regularly over the River Plate area except during disturbed conditions. Sea breezes are most marked during the summer. A land breeze (terral) can also be observed on quiet mornings. Mean cloud coverage is ≅ 4 oktas in the winter and 6 oktas in summer. The occurrence of maximum low cloud in the winter months is largely due to the greater frequency of incursions of polar maritime (Atlantic) air. Extensive cloud cover with a base at or below 1000 ft covers the area, and can extend as far as the Andean mountain range.

Cairo   With a north-easterly wind, fog may occur in any month but this usually disperses by 1000 hours. Low stratus in the early mornings from May to September is common. From December to May in association with southerly gale-force winds and sandstorms can be expected.

Calcultta (Dum Dum)   From November to May the north-east monsoon affects Calcutta leading to fog early in the morning. Between the north-east and south-west monsoon seasons a strong northwesterly wind can bring squalls in the evenings. During the late winter and early hot season, low stratus with a base of 200 to 800 ft occurs with moderate or strong southerly winds. During the hot season from March to June, dust haze appears with west to north-west winds. The nor'westers occur during this season, and winds can reach 50 kt or more.

Dakar   During summer, the ITCZ is very close and often passes just to the north. The weather is therefore wet and thundery during the months of July to September. The rest of the year is very dry and there is little cloud. Visibility is generally good, but the Harmattan can reduce visibility with dust haze.

Darwin   May to October the warm south-east trade winds of 25 kt bringing dust off the interior can reduce the visibility to 2000 m. November to April, the light north-westerly trade winds and the wet season bring heavy rain and squalls. Tropical cyclones can occur from December to April.

Delhi   During the north-east monsoon period, long spells of clear weather may give way to rain showers. The south-west monsoon brings in frequent Cbs.

Düsseldorf   The airport receives its bad weather from the NW. Winds can be directed through the Rhine valley from the SSE. Smoke from the Ruhr industrial area when brought in by light NE winds causes reduced visibility to be an aviation hazard. Westerly winds can produce fog if they are light, i.e. associated with anticyclonic conditions.

Gibraltar   The Rock of Gibraltar is 1300 ft AMSL. The airport is positioned to the north of the rock. If the wind is SE to SW, the turbulence produced makes landing very difficult, and is one of the recognised hazards for this airport. Turbulence can also be experienced with a south to south-west sea breeze particularly from April to September. When the wind is easterly (the *levanter*), as it is most frequently from July to October, 'banner' cloud is formed streaming from the peak of the rock. Fog is frequent during this period particularly around dawn and dusk. Fog is most frequent when the wind is a light north-easterly.

Greenland (Narsarssuak)   The airport is situated at the head of a fjord. The winds are aligned with the fjord and blow either SW or NE. During the winter, the winds are usually NE with the passage of depressions passing to the south of Greenland. The wind can be strong, over 20 kt and even gale force; high winds can close the airport. In spring with high pressure over the ice-cap, the winds will be NE. This complicates taking off, as the runway can only be used taking off to the SW. In the summer, the air mass is stable, but low stratus off the sea can also close the airport.

Harare   The ITCZ is at this location in the summer. Frequent Cbs with occasional early morning fog. In winter, a south-easterly wind brings in the *'guti'* resulting in low stratus and drizzle. The onset of the guti can be sudden.

Hong Kong   January to April is the season for the '*crachin*' when there is considerable low stratus and drizzle, and also sea fog. October to December is the time of the north-east monsoon which is mainly dry with low cloud; from May to September is the period of the south-west monsoon, which gives fair periods between heavy showers and thunderstorms. There are typhoons June to November.

Iceland (Keflavik)   The airport is exposed to Atlantic depressions from the sea from the SW through NW. The main problems are low cloud bases during frontal passage, and precipitation of rain and drizzle with snow in the winter. Squally gales can affect the airport thoughout the year. In winter drifting snow is a further hazard.

Johannesburg   Wet and thundery with thunderstorms coming in from the north-west October to March when low cloud can be expected; outside the showers, there is a possibility of fog but this clears early. Fog may occur on a few days each month throughout the wet season, again with a north-west wind. Fog occurs also in winter with easterly winds and at this time visibility is often reduced by smoke from the south-east especially in the mornings. The ITCZ does not reach this far south but its effects are noticeable.

Jakarta   From October to March the WSW monsoon prevails. There are frequent storms with showers, but visibility is good outside showers. The ESE monsoon prevails from April to September. There is less rain in total, but showers exist in the area. There is some haze, and surface winds hardly exceed 15 kt. There are marked sea breezes from the north most afternoons.

Karachi   During the SW monsoon (July to August) heavy showers and thunderstorms affect the airport. Rarely are these tropical storms. From October to March, early morning fog on occasions, but this clears quickly just after sunrise. Haze is frequent in the summer, and with winds exceeding 20 kt or so considerable dust is raised.

Kingston Jamaica   Generally a good weather record for aviation. Cloud bases rarely below 1000 ft. There is no fog. Thunderstorms formed by convection, maximum in December, when the cloud base may go down to 600 ft. The winter is the dry season. Hurricanes in the area during hurricane season, June to November.

Labrador (Goose) Worst conditions can be expected in winter with E or NE winds, when snowstorms can lower cloud base. This is associated with the passage of depressions south of the airport, or a high-pressure area to the north or north-west. During summer, an easterly wind from the Davis Strait, or the passage of a depression, can lower the cloud base to 300 ft. Fog and low cloud bases are, however, infrequent.

London Displays a wide variation of weather having a disturbed temperate climate. The general pattern is for summers to be warm and wet, and the winters to be cool and wet. The weather conditions vary widely from day to day and from region to region. London and the south-east of England enjoy more periods of sunshine and are drier and warmer than the north of England. Fog is most frequent from October to March and is often persistent; November is the worst month. Cloud bases can be very low in any season during a frontal passage, but particularly so in winter.

Melbourne Essendon Light winds from north bring fog in winter. Light winds from south-east bring industrial haze in summer and there is low stratus with southerly winds. Cold fronts from the south or west bring prolonged bad weather.

Montivideo (Carrasco) In spring and summer there are squalls resulting from the passage of cold fronts, although cold fronts from the SW can pass through in any season. During winter, fog or low stratus occurs with the passage of warm fronts to the north. With a depression centred to the north of the airport, gales from the SE develop, bringing in low cloud and precipitation, which may last for several days. Fog or low stratus can form in light westerly winds passing over the swamp areas to the west of the airport. Weak or occluded fronts can regenerate passing over the estuary of the River Plate. Thunderstorms are frequent during summer.

Montreal (Dorval) Winter is the worst time for the passage of depressions. From November to April there are frequent snowstorms with low cloud bases and poor visibility. Radiation fog is rare, but is experienced during the autumn. Smoke from Montreal is a further hazard in winter.

Munich (Reim) The airport lies just to the north of the Alps, and is thus sheltered from winds which are SE or SW. In winter, cloud bases of 400 to 500 ft can be expected in anticyclonic conditions, with light winds from other directions. Sudden fog occurs in winter in anticyclonic conditions and northerly winds, but often clears later at night when the winds back towards the south. With low pressure over Hungary during the winter and early spring, low cloud bases with precipitation can be expected. Visibility can be reduced to 2 to 6 km in haze from Munich in NW winds.

Nairobi There are two rainy seasons: from March to May, the '*long rains*', and from November to early January, the '*short rains*', when low stratus and fog is expected in the early hours after night rain. Afternoon thunderstorms may occur any time of year, but less frequently from June to September.

Newfoundland (Gander) The airport can experience advective fog from the sea in easterly winds, particularly during spring. However, fog and very low cloud can be expected in any season. In winter with polar continental air, the visibility can be very good. Snowstorms are frequent. October to April, gales are frequent. With westerly winds in winter, drifting snow is a hazard, with reduced visibility.

New York   New York (John F. Kennedy) is located on the south shore of Long Island about 10 miles south-east of the centre of New York city. The airport is situated on the north-east shore of Jamaica Bay, and the Atlantic Ocean lies a few kilometres to the south across the shallow bay. The surrounding terrain is quite flat, the nearest hills are $\cong$ 75 km to the north-west in New Jersey where the elevation is over 1600 ft. The average annual temperature is 12°C. The warmest month is July with an average maximum of 29°C. The coldest month is February with an average minimum of -4°C. Pure radiation fog is quite rare, but advection fog, sea fog and warm frontal fogs are not uncommon. May is the worst month for sea fog; advection fog and frontal fog are most common in the winter season. Frontal activity is vigorous in this area in all seasons, with a serious reduction of visibility at the passage of any warm fronts. During anticyclonic conditions in the autumn and winter, apart from the very occasional radiation fog, the visibility is good. Thunderstorms are mainly a summer phenomenon. Kennedy comes under the circulation of one or two hurricanes each year as they move up the eastern seaboard offshore of New England. An occasional hurricane may pass close enough to the airport to produce hurricane-force winds. Surface winds at Kennedy are quite variable with south-west winds predominating, but only on 18% of occurrences. Average speeds $\cong$ 13 kt, but speeds over 30 kt comprise $\cong$ 1% of the total. Calm winds are also uncommon comprising $\cong$ 2% of the total.

New Zealand (Christchurch)   Affected by the passage of depressions throughout the year, travelling west to east. The effects are diminished by the mountain chain to the west. Fog is a regular feature, and cloud bases are often below 600 ft. 80% of fogs are brought in from the sea, and may persist until midday. Rainfall is similar to London, and there is snow, mainly inland during the winter.

Rangoon (Migalodon)   October to April, the dry season, occasional early morning fog or low stratus, but soon disperses just after sunrise. During the wet season, May to early October, cloud bases can be below 1000 ft in heavy monsoon clouds.

Rome   Weather is good in summer except for occasional thunderstorms. At other times with a low pressure area in the Gulf of Genoa, the *sirocco* (from the SE) may bring cloud and rain. The '*libeccio*' is a strong squally wind from the SW. The *tramontana* is a NE wind and is associated with fine weather lasting for 3 or 4 days.

San Francisco   A range of coastal hills with elevations of 700 to 1900 ft extends from north-west to south about 7 km west of the airport. A broad gap towards the north-west between the San Bruno mountain and the coastal hills permits a considerable flow of maritime air from the Pacific Ocean to pass directly over the airport much of the time, particularly during the summer. The weather of the San Franciso Bay area can be divided into two separate and quite distinct seasons. The 'rainy' season (winter) from October to early May, associated with cyclonic and frontal conditions, and the 'dry or stratus' season from late May to October. Average annual rainfall is 450 mm. 77% of rainfall falls between December and March. Severe winter storms may be accompanied by strong winds, heavy rain and reduced visibility. During periods of calm, or near calm, and with a strong temperature inversion, the visibility may be reduced to zero in fog. Thick fog is reported 22 days a year, averaging 4 days a month in winter. Thunderstorms are rare, but do occur once or twice a year mainly in winter. The stratus season (summer) reaches its peak in August. This is persistent stratus or sea fog, which moves in during the night; it may produce drizzle. It often disappears after midday. Fog is rare during summer, but visibility may be reduced to below 5–6 km in haze and smoke. Mean surface temperatures (minimum in winter) ≅ 7°C. On occasions it can be lower. Summer temperatures are moderated by the morning low cloud cover, and in the afternoon by the strengthening sea breeze. Daily temperature average (maximum) 18° to 21°C (May through August). Surface winds during winter are south-westerly, 6 to 12 kt, although they can vary considerably during frontal passage, when they can exceed 50 kt and are from a south or south-westerly direction. Average winds in summer vary from calm to 10 kt in the mornings, rising to 20 to 30 kt midday and afternoon. 45 kt has been recorded. Direction is west or north-westerly, with little variation noted.

Santiago   Maximum rainfall west of the Andes occurs in winter, 81% at Santiago; winter is April to September. Cloud amounts depend on the source of the air mass. Northerly air flow produces wide cloud coverage of St and Sc from the Andes to the coast, rarely breaking during the day. More commonly, if the wind is variable, the air mass will stagnate and subsidence resulting from the proximity of the oceanic high-pressure area reduces the cloud amounts over Santiago to very little or nil. St and Sc are likely near the coast. Radiation fog may occur in the latter part of the night and early morning. Summer is October to March with very little rainfall at this time of year. Subsidence takes place in the air mass moving northwards over the Pacific and rainfall from associated cold fronts is very light. Sc forms at night on the coast with a base ≅ 1000 ft, tops ≅ 2000 to 3000 ft when tropical maritime air covers the area. It spreads inland at night, and may reach Santiago. It rapidly disperses after dawn, but could persist at the coast. There is very little or no medium or high cloud.

Singapore   Scattered showers or thunderstorms occur one day in two throughout the year. Rain mainly occurs in association with the north-east monsoon, November to March, with a late afternoon peak. *Sumatra* storms can affect Singapore during the night, around dawn or during the morning, with the south-west monsoon from June to September.

Sudan (Khartoum)   Sandstorms and duststorms are a problem. In winter, visibility may be reduced for 24 hours due to strong winds blowing sand from the north. In summer there are the *Haboobs* – severe duststorms associated with the ITCZ, particularly near Khartoum (May to September) where Cb clouds can lift dust up to great heights. Once the duststorm has passed there will be a torrential downpour and thunderstorms before conditions improve.

Sydney (Mascot)   During November to April, there are numerous cold fronts with Cbs and line squalls, the latter known as 'southerly busters'. North-westerly winds bring in dust and reduce the visibility. In winter there are less active fronts but early morning fog is common. With strong SW winds, turbulence is most severe from the winds passing over the high ground to the east and south.

Tokyo   Visibility is often poor because of smoke from local industry in winter and fog during the summer months. In winter, snow is a possibility. Sometimes typhoons reach Japan during June to November.

Trinidad (Piarco)   The wet season extends from June to September. Storms are mainly convective. Thunderstorms are most frequent during August to September. The rest of the year has good weather with lowest cloud base about 1000 ft, visibility general good and no fog.

Glossary

Ageostrophic wind   (or geostrophic departure) A vector representing the difference between the real wind and the geostrophic wind. Sometimes called geostrophic deviation.

Alaskan current   A current that flows north-west and westwards along the coasts of Canada and Alaska to the Aleutian Islands.

Alberta low   A low centred on the eastern slopes of the Canadian Rockies in the province of Alberta.

Anafront   A front at which the warm air is ascending the frontal surface to high altitudes.

Angular wave number   The number of waves of a given wavelength required to encircle the Earth at the latitude of the disturbance. Also known as hemispherical wave number.

Antarctic anticyclone   The glacial anticyclone which has been said to overlie the continent of Antarctica. Analogous to the Greenland anticyclone.

Antarctic convergence   The ocean polar front indicating the boundary between the subantarctic and subtropical waters. Also known as the southern polar front.

Antarctic front   The semi-permanent, semi-continuous front between the Antarctic air of the Antarctic continent and the polar air of the southern oceans. Generally compared to the Arctic front of the northern hemisphere.

Antitrades   A deep layer of westerly winds in the troposphere above the surface trade winds in the tropics.

Arctic front   The semi-permanent, semi-continuous front between the deep cold Arctic air and the shallower, basically less cold polar air of northern latitudes.

Arctic haze   A condition of reduced horizontal and slant visibility (but unimpeded vertical visibility) encountered by aircraft flying at 30 000 ft or more over Arctic regions.

Arctic high   A weak high that appears on mean charts of sea-level pressure over the Arctic basin during spring, summer and early autumn. Also called an Arctic anticyclone, polar anticyclone or polar high.

Arctic mist   A mist consisting of ice crystals; a very light ice fog. Free air temperature below $-30°$C.

Arctic sea smoke   Steam fog, but often specifically applied to steam fog rising from small areas of open water within sea ice.

Bai   A yellow mist prevalent in China and Japan in spring and autumn, when the loose surface of the interior of China is churned up by the wind. Clouds of dust rise to great heights and are carried eastwards, where (acting as precipitant nuclei) they fall as a yellow mist.

Barotropic and baroclinic atmospheres   When neither the direction nor the speed of the wind changes in the vertical plane within the troposphere then the atmosphere is said to be *barotropic*. Although this condition never actually exists in the free atmosphere, the tropical regions often come close to it. If the wind changes in direction and/or speed in the vertical plane within the troposphere, then the atmosphere is said to be *baroclinic*. This condition is a common phenomenon throughout the world, but reaches its maximum in the mid-latitudes. The development and movement of pressure systems are directly related to the degree of baroclinicity in the atmosphere.

Brazil current   The warm ocean current that flows southwards along the Brazilian coast below Natal. The western boundary current in the South Atlantic Ocean.

Californian Current   The ocean current flowing southwards along the western coast of the United States to northern Baja California.

Californian fog   A fog peculiar to the coast of California and its coastal valleys. Off the coast, winds displace warm surface water, causing colder water to rise from below, and results in the formation of advective fog. In the coastal valleys, fog is formed when moist air is blown inland during the afternoon and is cooled by radiation during the night.

Canary current   The prevailing southward flow of water along the north-west coast of Africa.

Cape Horn current   That part of the west wind drift flowing eastwards in the immediate vicinity of Cape Horn, then curving north-eastwards to continue as the Falkland current.

CAVOK   A code used in terminal aerodrome forecasts (TAFs) to indicate when the following conditions occur simultaneously at the time of observation. Visibility 10 km or more, no cloud below 1500 m (5000 ft) or below the highest minimum sector altitude, whichever is greater, no cumulonimbus, no significant weather phenomena at or in the vicinity of the aerodrome.

Cold high   At a given level in the atmosphere, any low that is generally characterised by colder air near its centre than around its periphery. Also known as a cold anti-cyclone, or cold-core high.

Cold low   At a given level in the atmosphere, any low that is generally characterised by colder air near its centre than around its periphery. Also known as a cold-core cyclone, or a cold-core low.

Cold pool   A closed centre of low thickness on a thickness chart. This indicates low mean temperature of the air within standard pressure levels. (Usually 1000 mb and 500 mb pressure levels). The term is usually applied to cold air of appreciable vertical extent that has been isolated in lower latitudes as part of the formation of a cut-off low. Also known as a cold-air drop or cold drop.

Cold tongue   In synoptic meteorology, a pronounced equatorward extension or protrusion of cold air.

Colorado low   A low which makes its first appearance as a definite centre in the vicinity of Colorado on the eastern slopes of the Rocky Mountains. Analogous to the Alberta low.

Convergence zone   An area of net flux of air horizontally. When convergence occurs at low levels this influx must be compensated by upward motion. The intertropical convergence zone (ITCZ) is an example, where the trade winds of the two hemispheres meet.

Cotton belt climate   A type of warm climate characterised by dry winters and rainy summers, that is a monsoon climate, in contrast to a Mediterranean climate.

Crachin   A period of light rain accompanied by low stratus clouds and poor visibility which frequently occurs along the China coast and adjacent sea in January to April.

Cyclogenesis   The initiation of a secondary low or strengthening of cyclonic circulation around an existing cyclone or depression.

Cyclolysis   The weakening or disappearance of cyclonic circulation around an existing cyclone or depression.

Cyclostrophic wind   The horizontal wind velocity where the centripetal acceleration balances the horizontal pressure gradient.

Darling shower   A duststorm caused by cyclonic winds in the vicinity of the River Darling in New South Wales, Australia.

Density current   An intrusion of a dense air mass beneath a lighter air mass. This describes the mode of action of a cold front.

Diamond dust   The description of precipitation of very small (unbranched) ice crystals forming in air supersaturated with respect to ice at temperatures below $-30°C$. Diamond dust accounts for the major accumulation of 'snowfall' in the interior of Antarctica.

Doister   In Scotland, a severe storm from the sea. Also know as a deaister, or dyster.

Doldrums   The equatorial oceanic regions of light variable winds (mainly westerly). Accompanied by heavy rains, thunderstorms and squalls. These belts have a north and south movement, following the sun's latitude but lagging by some 6 to 8 weeks. The movement is some 5° either side of the mean position. They are, however, very variable in position and extent.

Downrush   A term sometimes applied to the strong downward flowing air current associated with the dissipating stage of a thunderstorm.

Drops and droplets   Drops are usually water drops greater than 0.2 mm (200 µm) in diameter. Droplets are water drops smaller than 0.2 mm. The largest raindrops are in the order of 5 mm (exceptionally 8 mm), but turbulence in cloud constantly shatters the drops, particularly at diameters of $> 5\,mm$. (Raindrops are usually $0.5\,mm$ ($500\,\mu m$) and drizzle $> 0.5\,mm$, cloud droplets are $< 0.1\,mm$.)

Dry climate   In W Köppens climatic classification, the category which includes steppe climate and desert climate and is defined strictly by the amount of annual precipitation, seasonal distribution and annual temperature. In C W Thornwaite's climatic classification, a dry climate is described as 'having a moisture index of less than zero'. It includes the dry arid, semi-arid and sub-humid climates.

Dry spell   Usually a condition of abnormally dry weather, less severe than a drought. In the USA, a dry spell is a period where no measurable precipitation has been recorded for a period of not less than two weeks.

Dry tongue   An intrusion of relatively dry air into a region of air with much greater moisture content.

Easterly wave   A migratory low pressure trough with local intensification of cloud and precipitation. Occurs in the Caribbean and other tropical regions.

Entrance region   The region of confluence at the upwind end of a jet stream.

Equatorial easterlies   During the summer in the appropriate hemisphere, the trade winds can extend up to 10 km. They are generally not topped by upper westerlies. If, however, westerlies are present, they are very weak and shallow, and have no significant effect on the weather. The trade winds are described as deep easterlies, or deep trades.

Equatorial westerlies   The westerly winds occasionally found in the Equatorial trough, and separated from the mid-latitude westerlies by the broad belt of easterly trade winds.

Equinoctial rains   These are rainy seasons which occur regularly at or shortly after the equinoxes in many areas of the world.

Equiparte   In Mexico, during October to January, heavy cold rains that last for several days. Also known as equipatos.

Exit region   The region of difluence at the downwind end of a jet stream.

Forty saints storm   A southerly gale in Greece. It occurs just before the equinox in March.

Frontal cyclone   Any cyclone associated with a front. Often used synonymously with extratropical cyclone or wave cyclone, as opposed to a tropical cyclone which is non-frontal.

Frontal wave   A horizontal wavelike deformation of a front, restricted to the lower levels. It coincides with the maximum of cyclonic circulation in the adjacent flow. This mechanism could lead to the formation of a wave cyclone.

Frontogenesis   The development or intensification of a front.

Frontolysis   The weakening or disappearance of a front.

Frost smoke   Fog which is formed in the same manner as steam fog but at lower temperatures. The suspensions consist of ice particles instead of water droplets.

Garua   A dense fog or drizzle from low stratus clouds on the west coast of South

America. It creates very cold conditions that may last for several weeks in winter. Also known as *camanchaca*.

Glacial anticyclone   A type of semi-permanent anticyclone found over the ice-caps of Greenland and Antarctica. Also known as a glacial high.

Great Basin high   A high pressure system centred over the Great Basin of the western United States. A common synoptic winter feature.

Guba   In New Guinea, a rain squall at sea.

Gully squall   A violent squall of wind from the mountain ravines on the Pacific side of Central America.

Guti   In Zimbabwe, a dense stratocumulus overcast, often reaching the ground. Drizzle most likely. Occurs in the early summer and associated with easterly winds that invade inland bringing in cool (stable) maritime air from the Indian Ocean.

Guttra   In Iran, sudden squalls, usually in May.

Haar   A sea fog or fine drizzle which drifts in from the sea in coastal areas of east Scotland (and NE England). Usually a summer phenomenon.

Horse latitudes   The description given to the belts of variable light winds at about 30–40° North and South latitudes. Fine weather prevails because of the association with the subtropical anticyclones. There is a movement north and south of the mean position, but only slightly.

Humidity mixing ratio   The ratio of the mass of water vapour to the mass of dry air. Usually expressed as grams of water vapour per kilogram of dry air.

Index of aridity   The measurement of the aridity of a region $= P/(T+10)$, where $P =$ annual precipitation in centimetres, and $T =$ annual mean temperature in degrees Celsius.

Isohyet   A line drawn on a chart joining places of equal rainfall amount.

Isopleth   (ISO) A prefix meaning 'equal'. Extensively used in meteorology to denote lines drawn on a map or chart to display the geographical distribution of any element, each line being drawn through places at which the element has the same value, e.g. isobar, isotherm. The words 'isogram' and 'isopleth' are used as generic names of this type.

Isothermal ribbon   A zone of crowded isotherms on a synoptic upper level chart. The temperature gradient would be many times more than normally encountered in the atmosphere.

Kai baisakhi   A short-lived dusty squall that occurs at the onset of the SW monsoon in April in Bangladesh.

Katafront   A front (usually a cold front) where warm air descends the frontal interface.

Local storm   A storm of mesometeorological scale. Therefore, thunderstorms, squalls and tornadoes are placed in this category.

Low index   A relatively low value of the zonal index which in middle latitudes indicates a relatively weak westerly component of wind flow, which implies a stronger north–south motion and the attendant weather that accompanies a low zonal index.

Macroclimate   The climate of a large geographic (continental size) area.

Meridional cell   This is generally a closed circulation in a vertical plane orientated along a geographic meridian. An example is the Hadley cell.

Meridional flow   Airflow in the direction of the geographic meridian. Flows that are north to south, or south to north. The accompanying zonal component is weaker than usual.

Mesoclimate   The climate of small areas of the Earth's surface, which may not represent the general climate of the district. Also, a climate characterised by moderate temperatures, that is in the range of 20 to 30°C. This is also known as a 'mesothermal' climate.

Mesoscale   The scale appropriate to atmospheric systems between individual cumulus clouds on the one hand and major depressions and anticyclones on the other, that is from about ten to a few hundred kilometres.

Microburst   A highly concentrated powerful downdraught: low level. Typically less than 5 km across, lasts from one to five minutes. Downdraught speeds up to 60 kt (sometimes more). Can be either wet or dry. *Wet*: low cloud base with rain or hail, very heavy. *Dry*: high cloud base.

Microclimate   The physical state of the atmosphere close to a very small area of the Earth's surface, often in relation to living matter such as crops or insects. In contrast to climate, microclimate generally pertains to a short period of time.

Mixing ratio   The ratio of one gas to the total mass of gases in the atmosphere.

Moisture content   Sometimes described as *specific humidity*, the ratio of the mass of water vapour to the mass of moist air. This is not the same as the mixing ratio, but the two values are nearly equal.

Nephanalysis   The analysis of cloud types and amounts obtained from satellite imagery. Usually displayed on a geographical map.

Octas or oktas   Quantification of cloud amounts in eighths.

Palouser   A duststorm of north west Labrador.

Periglacial climate   A climate which is characterised by the neighbouring outer perimeter of the ice-cap or continental glacier. The main characteristic of the weather is the frequent cold and dry winds off the ice. It is also the region where there seems to be frequent cyclonic activity.

Potential instability   Air which is initially stable but becomes unstable after being lifted. On an aerological diagram, the wet-bulb temperature decreases with height at a greater rate than the saturated adiabatic lapse rate through the lowest level, the air being moist below and dry at height. When potentially unstable air is forced to rise by frontal displacement, thundery rain can break out.

Rainfall regime   The characterisation of seasonal rainfall distribution at any place.

The main rainfall regimes, as defined by W G Kendrew, are equatorial, tropical, monsoonal, oceanic and continental westerlies, and Mediterranean.

Saffir–Simpson scale   A hurricane intensity scale used by the United States National Weather Service for assessing the damage that is likely from wind and storm surge from a hurricane. The scale ranges from 1 (minimal) to 5 (catastrophic).

Sandstorm   A strong wind > 16 kt that raises sand particulates. Diameters range between 0.8 and 1 mm, therefore it does not rise more than a few metres from the ground, rarely above 11 m. Dust particulates are very fine and can reach the tropopause in storms.

Savannah   The name given to a type of tropical climate. The characteristics are wet and dry seasons. The predominant vegetation is tall tropical grass (savannah).

Semi-permanent   The term 'semi-permanent' is used by meteorologists to describe features that are present on most maps over a season, i.e. the semi-permanent winter high pressure system over Asia.

Smog   Means smoke and fog combined. The smoke acts as condensation nuclei. Photochemical smog happens when slowly moving or stagnant air is reacting to sunlight on the hydrocarbon exhaust products from industry and motor cars that is suspended in the atmosphere. This produces 'ozone' among other oxidising agents. A phenomenom often seen in Los Angeles.

Smokes   Dense white haze and dust clouds common in the dry season on the Guinea coast of Africa. Particularly noticeable with the onset of the Harmattan.

Sonora   A summer thunderstorm in the mountains and deserts of southern California and Baja California.

Steam fog   Fog formed when water vapour is added to air which is much colder than the vapour source. Occurs when very cold air drifts across relatively warm water. It is, however, warmed at the sea surface in the first instance and is able to allow evaporation to take place. A little higher in the air, the temperature is lower than the surface dew-point condition, and precipitation takes place forming low-lying fog. Also called frost smoke, sea mist, sea smoke, steam mist and water smoke.

Taiga climate   A climate which produces taiga vegetation, i.e. it is too cold to produce significant tree growth, but milder than the tundra climate. It is moist enough to produce appreciable vegetation. It is more commonly known as a 'subarctic' climate.

Taino   A tropical cyclone in the Greater Antilles.

Tephigram   (Aerological diagram) A graphical representation of the observations of pressure, temperature and humidity made in a vertical sounding of the atmosphere.

Thickness   The *geopotential* height difference at a given place between specified pressure levels. Thickness values relating to selected standard pressure levels are obtained from radiosonde observations and are plotted on geographical 'thickness charts'.

Thickness lines   Contours are drawn, at an appropriate thickness interval (often between the 1000 mb and 500 mb layers) joining places of equal thickness. Thickness analysis has an important role in synoptic meteorology.

Tofan   A violent spring storm in the mountains of Indonesia.

Tundra   The name given to the treeless lands of northern Canada and Eurasia in the immediate neighbourhood of the Arctic Circle. The significance of the tundra areas is that the temperature is subzero for most of the year, and 2 to 3 months of the year in summer rises to positive values. However, only about one foot (30 cm) below the surface the temperature is subzero all the year round (permafrost).

Tundra climate   A climate that produces tundra type vegetation. It is too cold to produce tree growth, but does not have a permanent ice cover.

Turbonada   A short thunderstorm/squall, occurring on the northern Spanish coast, sometimes producing waterspouts.

Veranillo   The lesser dry season, on the Pacific coast of Mexico and Central America. It lasts only a few weeks. It is hot and dry and breaks up the summer rainy season.

Verano   The main dry season in Mexico and Central America. Generally November to April.

Water vapour absorption   The absorption of certain infrared wavelengths by atmospheric water vapour. A process of fundamental importance in the heat transfer of the atmosphere.

Wave clyclone   A cyclone that forms on a front. The cyclone centre deforms the front into a wave. Sometimes called a wave depression.

Westerlies   The dominant surface westerly atmospheric flow centred at the mid-latitudes of both hemispheres. The 'west wind belt' extends $\cong$ 35 to 65° latitude. Other names are: circumpolar westerlies, mid-latitude westerlies, middle-latitude westerlies, polar westerlies, sub-polar westerlies, subtropical westerlies, temperate westerlies, zonal westerlies, zonal winds.

Williwaw   A very violent storm and squall in the Straits of Magellan. It is most frequent in the winter, but can occur in any month.

Wind shear   A change in any of the components of wind vectors in a given direction, horizontal or vertical.

WMO (World Meteorological Organization)   The WMO is a specialised agency of the United Nations. It specialises in the field of meteorology. Its permanent Secretariat is based in Geneva.

World area forecast centre (WAFC)   A meteorological centre designed to prepare and supply upper-air forecasts in digital form on a global basis to regional area forecast centres.

World area forecast system (WAFS)   A worldwide system by which world and regional area forecast centres provide aeronautical meteorological en-route forecasts in uniform standard formats.

Xeothermic   Characterised by dryness and heat.

Zonal flow This is generally west–east flow. Counter flow, or east–west flow, is referred to as a negative zonal flow.

Index